LOW ATTAINERS IN PRIMARY MATHEMATICS

LOW ATTAINERS IN PRIMARY MATHEMATICS

Editors

Berinderjeet Kaur • Masura Ghani

National Institute of Education, Singapore

World Scientific

NEW JERSEY • LONDON • SINGAPORE • BEIJING • SHANGHAI • HONG KONG • TAIPEI • CHENNAI

Published by

World Scientific Publishing Co. Pte. Ltd.
5 Toh Tuck Link, Singapore 596224
USA office: 27 Warren Street, Suite 401-402, Hackensack, NJ 07601
UK office: 57 Shelton Street, Covent Garden, London WC2H 9HE

British Library Cataloguing-in-Publication Data
A catalogue record for this book is available from the British Library.

LOW ATTAINERS IN PRIMARY MATHEMATICS

ISBN-13 978-981-4374-87-3
ISBN-10 981-4374-87-3

Printed in Singapore by B & Jo Enterprise Pte Ltd

Preface

This book, comprising five chapters, is based on a funded study of the Office of Education Research at the National Institute of Education in Singapore. The study explored low attainers in primary mathematics, specifically primary four pupils from nine schools across Singapore. The impetus for the study was the growing proportion of low attainers in primary mathematics that was evident from both international and national studies data.

The first chapter of the book introduces the study. In the next three chapters, part of the data of the study is reported, namely:

- the mathematics content knowledge and skills of the pupils in the study;
- characteristics of the pupils related to school, home and self;
- the mathematics learning experiences of the pupils.

The last chapter of the book puts forth recommendations for teachers arising from the findings in chapters 2, 3 and 4.

This peer-reviewed book contributes towards the professional development of teachers in Singapore and also research literature on low attainers in primary mathematics. Lastly we like to dedicate this book to all the pupils and teachers who participated in the study.

Berinderjeet Kaur, Koay Phong Lee, Foong Pui Yee, Chang Suo Hui, Masura Ghani, Akhila Sudarshan.

September 2011

Contents

Chapter 1

An Exploratory Study on Low Attainers in Primary Mathematics (LAPM)

Berinderjeet KAUR KOAY Phong Lee

FOONG Pui Yee Akhila SUDARSHAN

This introductory chapter provides the background to an exploratory study on Low Attainers in Primary Mathematics (LAPM) which was carried out in Singapore. It provides the rationale for the study and the definition of low attainers that guided the study. It also reviews literature on low attainers in mathematics, and states the research questions which were investigated in the study. Lastly, the method and instruments used for data collection in the study are described.

1 Why Study Low Attainers in Primary Mathematics

Ginsburg, Cooke, Leinward, Noell and Pollock (2005) found in their secondary analysis of the Trends in International Mathematics and Science Study (TIMSS) 2003 data for the US that pupils who were failing at grade 4 continued to fail at grade 8 subsequently. Table 1 shows the TIMSS 2007 mathematics mean score by the percentiles of the participants from the top four performing countries for grade 4 mathematics. From the table it is apparent that grade 4 pupils from Singapore were underachieving, in mathematics, compared to their peers from Hong Kong and Chinese Taipei when comparing their mean scores of up to the 5th, 10th, 15th, and 20th percentiles. Such findings from international studies and also national examinations have prompted researchers and policy makers to explore the needs of and provisions for low attainers in primary schools.

Table 1
TIMSS 2007 grade 4 mathematics mean scores

Country	Rank	TIMSS 2007 Grade 4 mathematics mean scores of pupils up to the nth percentiles				
	n =	5th	10th	15th	20th	25th
Hong Kong	1	462	488	503	514	524
Singapore	2	408	440	461	477	490
Chinese Taipei	3	427	453	469	481	490
Japan	4	404	432	449	463	474

Only a few studies done so far in Singapore provide findings that may relate to low attainers in primary mathematics (Loo & Fong, 1996; Foong, 1999; Lee, 1999; Leong, 1999; Lim, 1999; Ng, 2003). Loo and Fong (1996) found that both mathematical and psychological factors correlated with mathematics achievement while Ng (2003) found that social economic status related variables also correlated with mathematics achievement. Foong (1999) found that low attainers were diverse in their needs and they were heterogeneous. Lim (1999) found that low attainers preferred formally designed learning environments and Lee (1999) noted that academic achievement scores related positively with subject-related self-esteem. Leong (1999) in her study on parental involvement and effects on the academic performance of students found that there was no conclusive evidence to suggest that the higher the degree of parental involvement, the better the academic performance of the student. Though these studies do provide glimpses of the characteristics of low attainers, none of them have investigated the content knowledge of low attainers, their learning experiences and how schools organise, motivate and inspire them.

In Singapore mathematics is a core subject in the school curriculum for primary and secondary education. Failure in mathematics due to factors that may be controlled for would be unjust for the pupils. Therefore this study hopes to shed light on the needs of low attainers in primary mathematics and help them achieve their potential.

2 Who are the Low Attainers in Primary Mathematics

Early research on low attainment in mathematics describes low attainers in mathematics to be those pupils outside special schools who fall, for whatever reason, into the bottom 20 per cent of mathematical attainment in their age group in national assessment (Denvir, Stolz, & Brown, 1982). Denvir, et al., (1982) have used the term low attainer to describe the observable performance of a pupil, without implying a cause. Other labels attached to such performing students are slow learners, at risk students, special needs children, less able children and so on. Realising that these labels point towards deficiencies on the part of a learner, Haylock (1991) adopted a pragmatic approach and used the term "low attainers" to define pupils who attain very much less in mathematics when compared to their contemporaries. The use of this term does not make any judgment about the reasons for low attainment in mathematics.

Like Haylock (1991), in this study we too have chosen to adopt a very pragmatic approach by using the term "low attainers" to be all encompassing of what some may label as: less able children, slow learners and under-achievers.

3 Review of Literature

Boaler (1997) in her work found that teachers commonly believe that children in low attaining groups are similar. This is contrary to research on low attainers in mathematics which has found that low attainers are heterogeneous (Foong, 1999; Houssart, 2004). Descriptions and explanations of low attaining pupils have tended to emphasise their diversity. For example, Haylock (1991) talks about a broad range of and great variety in differences related to children with low attainment in mathematics. He used case studies of individuals, to demonstrate complex combinations of difficulties in understanding and engagement with mathematics by low attainers. Similarly, Denvir, et al., (1982) also draw on examples to illustrate the diversity of low attainers. Both Haylock (1991) and Denvir, et al. (1982) assert that these children will not form a homogenous group and some of the causes of their low attainment may be school related while others may not be and hence are beyond the control of the school.

Haylock (1991) in his work suggested that a "*deficit model*" of low attainment that emphasises mainly the diagnosis of the child's difficulties in understanding and mastering mathematics was an inadequate way of analysing the problem. He proposed the "*ecological model*" which attempts to in addition describe the relationships between the child and the whole learning environment to be a better one as it is more likely to produce useful insights. According to Haylock (1991), at least three factors may be related to low attainment in primary mathematics. The factors are as follows:

- Some significant characteristics of school mathematics, such as
  - accuracy and concentration,
  - a symbolic language,
  - abstract concepts and relationships, etc.
- Some specific intellectual or behavioural characteristics frequently associated with low attainers, such as
  - reading and language problems,
  - perceptual problems (spatial orientation, etc.),
  - social problems,
  - mathematics anxiety, etc.
- Some of the shortcomings in the way the subject is often taught and assessed such as
  - the preoccupation of "completing the syllabuses",
  - meaningless routines,
  - the teacher's way is the only way,
  - practice and yet more practice,
  - assessment tasks are complex and tedious, correlate negatively with motivating for pupils

Mercer and Mercer (2005) stated that pupils failing in mathematics have one or more of the following main difficulties:

- Intellectual problems on the part of the pupil, such as
  - short attention span,
  - low retention (forgets facts and skills),
  - cognitive immaturity,
  - slow in grasping concepts,

  - problem in organising knowledge and experience,
  - inaccurate computation / procedure,
  - lack of verbal, spatial ability.
- Affective, such as
  - negative attitude and low motivation,
  - anxiety / panic,
  - low self-concept.
- Pupil's educational experience in mathematics, such as
  - lack of appropriate background knowledge,
  - inadequate concrete practical experience,
  - symbols too abstract,
  - lack of pre-requisite knowledge,
  - poor mathematical language,
  - use of rules without understanding.

Hence, any combinations of the above may result in low attainment in primary mathematics. Also, Loo and Fong (1996) in their study on variables associated with mathematics achievement of Primary 5 pupils in Singapore found that although both mathematical and psychological factors were related to mathematics achievement, mathematical factors contributed a higher variance to mathematics achievement than psychological factors. In their study mathematical factors comprised logical thinking ability, computational skills, mathematical language, and mathematical concept while psychological factors comprised attitude towards mathematics, interest in mathematics and memory.

Pupils' experiences within school grouping processes and practices are highly significant to their learner identities, their expectations, aspirations and motivations (Dunne, Humphreys, Sebba, Dyson, Gallannaugh, & Muijs, 2007). While prior performance data inform this process, teachers inevitably play a significant part in grouping decisions, which in turn influence their pedagogical approaches with low attainment groups (Kutnick, et al., 2006). In this sense, the organisational structures and processes in schools are highly pertinent to the pedagogical context for low attaining pupils and their teachers.

Attainment grouping in practice appears to advantage some pupils while disadvantage others. Ireson, Hallam, Hack, Clark, and Plewis (2002) found that pupils with high attainment in mathematics benefitted from homogeneous attainment groups while those with low attainment in mathematics benefited from mixed attainment groups. Hallam and Ireson (2007) found that teachers preferred homogenous attainment groups as it allowed pupils to learn at the appropriate level and pace for the group. Slavin (1990) suggested that grouping by attainment increases the motivation of the lower attaining pupils by removing any competition or intimidation from working alongside higher attainers but a later research by Boaler, William, and Brown (2000) does not support this suggestion. The lack of clear evidence to support one form of grouping may reflect the greater influence that other factors have on outcomes such as effectiveness of teaching (Day, Stobart, Sammons, Kington, Gu, Smees, & Mujtaba, 2006) and curriculum offered (Kutnick, Blatchford, & Baines, 2002; Blatchford, Kutnick, Baines, & Galton, 2003)

It is well known that pupils learn more when the school and classroom environments are positive and supportive (Christenson, Ysseldyke, & Thurlow, 1989). The teacher is key, as he or she arranges the physical (seating) and academic (e.g. scheduling, method of lesson presentation, nature of assessment, homework, etc.) variables and establishes the affective nature (e.g. encouragement, competitiveness, and cooperation) of the classroom. Teacher expectations, encouragements, evaluations, attentiveness, and attitudes greatly influence pupils' perceptions of themselves as learners (Mercer & Mercer, 2005). Because many low attainers in primary mathematics have negative perceptions of their ability, teachers must create and maintain a supportive environment. Lee (1999) in her study on low attainers in primary mathematics in Singapore notes that one way to help low attainers is by enhancing their mathematics specific self-esteem through creative and varied teaching strategies.

Sprick (1985) notes that learning is greater and behaviour is more appropriate in classrooms where teachers attend to positive events more than to negative events. Borich (1992) and Brigham, Scruggs, and Mastropieri, (1992) maintain that teacher enthusiasm is an important aspect of teacher effort, is positively correlated to pupil achievement, and helps establish a positive and supportive learning environment.

Alderman (1990) noted that a positive learning environment and pupil learning are enhanced when teachers believe that all pupils can learn and that teachers can make a difference. It is also equally important to provide the right kinds of learning experiences considering pupil preference for learning styles. Lim (1999) in her study on preferred learning styles of low achieving and high achieving pupils in a Singapore primary school found that low achievers preferred formally designed learning environments. Christenson, et al., (1989) state that a positive learning environment is built upon the use of realistic expectations of pupil learning, the development of instructional plans that consider pupil characteristics and needs, the use of reinforcement for pupil productivity, the use of active monitoring of pupil progress, and the belief that all pupils will experience academic success. Pupils need to be motivated and engaged in the learning process. Often low attainers in mathematics lose their motivation for learning due to repeated failure at mathematics tests. Setting realistic instructional goals and determining mastery criteria are important to pupil motivation. To experience success and link it to one's own effort are critical for the development of motivation in pupils (Alderman, 1990). Hence, opportunities to experience such successes must be available for low attainers in primary mathematics.

From the factors that may be related to low attainment in primary mathematics (Haylock, 1991) and the difficulties that may be experienced by low attainers in mathematics (Mercer & Mercer, 2005), it is apparent that teachers working with low attainers would need customised approaches to help their pupils. Cockburn (1999), notes that difficulties with mathematics may stem from three perspectives: child (experience, expertise, mathematical knowledge and understanding, imagination and creativity, mood, attitude and confidence) task (mathematical complexity, presentational complexity, and translational complexity), and teacher (attitude and confidence, mood, imagination and creativity, knowledge, expertise and experience). There is an abundant of research on how specific causes of low attainment may be addressed (see Haylock, 1991; Houssart, 2004; Mercer & Mercer, 2005; & Cockburn, 1999).

Following on from the above review of relevant literature, it is apparent that an ecological model would be a preferred one to research the characteristics of low attainers in primary mathematics. The literature also highlights that there are a myriad of factors associated with low attainment in primary mathematics. From the focused discussions the research team held with heads of mathematics departments in primary schools, it was found that teachers do engage their low attainers in remedial lessons. This leads us to infer that there may be a misfit of needs and remediation. Hence an exploratory study of the education of low attainers in primary mathematics that focuses on the characteristics of low attainers and their education, in particular the practices and processes used by schools to identify and organise them, how teachers motivate and inspire them, and how schools and teachers address their diverse learning needs is necessary before any form of intervention strategies can be planned for levelling up the performance of low attainers.

4 The Research Questions

The review of literature led to a conceptual framework, shown in Figure 1, which was adopted by the study.

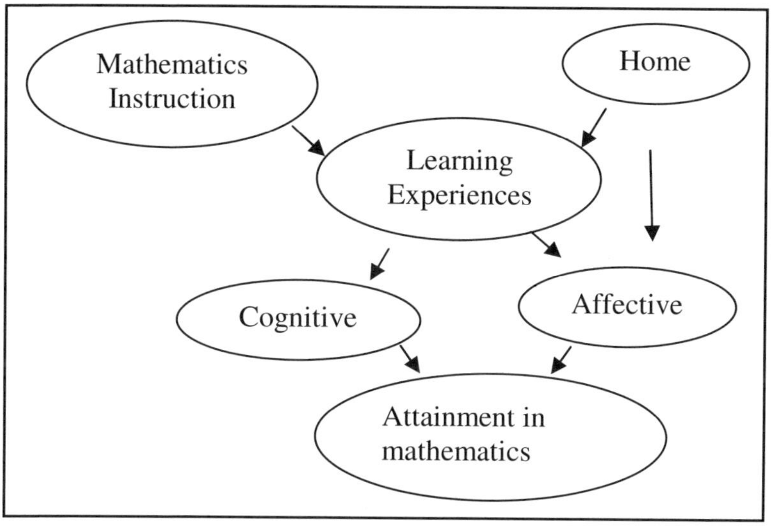

Figure 1. Conceptual framework for the study

The framework shows that mathematics instruction and the home impact pupils' learning experiences. These experiences shape pupil's cognitive and affective aspects of learning that inevitable affect their mathematics attainment. Drawing on the framework of the study, six research questions were formulated. The questions are as follows.

1. How did the low attaining pupils perform in mathematics tests that tested mastery of Primary 3 mathematics content knowledge and skills?
2. What were the characteristics of low attaining in mathematics pupils related to school (behaviour during class, absenteeism, interactions with peers), home (support, resources) and self?
3. What were the characteristics of the low attaining in mathematics pupils related to their experiences of learning mathematics?
4. What were the practices and processes used by schools to identify and organise low attainment groups?

5. How did schools and teachers motivate and inspire low attainers in primary mathematics to learn?
6. How did schools and teachers address the diverse learning needs of low attainers in primary mathematics?

In this book, we report the data and findings for only three of the six research questions, which are:

1. How did the low attaining pupils perform in mathematics tests that tested mastery of Primary 3 mathematics content knowledge and skills?
2. What were the characteristics of low attaining in mathematics pupils related to school (behaviour during class, absenteeism, interactions with peers), home (support, resources) and self ?
3. What were the characteristics of the low attaining in mathematics pupils related to their experiences of learning mathematics?

5 Method

Both qualitative and quantitative data were collected for the study. Four main methods were used to collect data. They were paper and pencil tests, surveys, interviews and observation. Data were collected from school leaders, heads of mathematics departments, mathematics teachers, pupils and their parents. In accordance with the ethics requirements of the study, consent was sought from all participants.

5.1 *The sample*

Nine government schools from across Singapore participated in the study. Participation by the schools was voluntary. The schools identified the pupils for the study, based on the school's criteria of low attainers. The parents of the pupils, identified by the schools, were asked for their consent regarding their child's participation in the study. Pupils' whose parents did not consent were excluded from participation. Pupils whose parents gave consent were asked for their assent. These pupils were the subjects of the study. Table 2 shows the number of pupil participants.

Table 2
Pupil participants of the study

School Code	Location	Number of pupils nominated for the study	Number of pupils with consent from parents	Number of pupils who took all the benchmark tests	Number of pupils who were interviewed
01	East	37	24	21	23
02	East	69	51	41	48
03	East	43	32	30	29
04	East	33	24	18	22
05	East	78	56	52	55
06	West	52	35	32	31
07	North	74	12	54	11
08	West	55	50	48	45
09	West	119	91	94	82
	Total	560	375	390*	346

* This number is larger than the "bona fide participants" of the study as in schools 7 and 9, the teachers administered the benchmark tests to a larger group of pupils in their respective schools.

While the study was in progress, four of the participants left their respective schools. Two went back to their home country and another two transferred to a school that was not in the study. Eventually, the study only managed to interview 346 pupils as 29 pupils were absent on the days of the interviews.

Table 3 shows the number of school leaders and teachers who participated in the study. There were altogether 9 school leaders and 23 teachers who participated. All the teachers were teaching the pupil participants.

Table 3
School leaders and teacher participants of the study

School Code	Location	Number of school leaders	Number of teachers
01	East	1	1
02	East	1	4
03	East	1	1
04	East	1	1
05	East	1	4
06	West	1	4
07	North	1	3
08	West	1	2
09	West	1	3
	Total	9	23

Table 4 shows the profile of the teacher participants of the study. More than 75% of the teachers were female. Almost half of the teachers were in the age range of 30 – 39 years, with the rest in equal proportions in the age ranges of less than 30 years and 40 – 49 years. More than 75% of the teachers had less than a year to 10 years of experience of teaching mathematics in primary schools. Almost 15% of the teachers were in their first year of teaching mathematics, i.e. beginning teachers and another 35% had 1 to 3 years of experience of teaching mathematics in a primary school. Only a fifth of the teachers had more than 10 years of teaching mathematics experience in primary schools. Majority (almost 80%) of the teachers had at most three years of experience teaching Primary 3 and / or Primary 4 mathematics. In addition to the three beginning teachers another 7 teachers were also teaching the low attainers for the first time. Majority of the rest (11 in total) had only taught low attainers for at most three years. All the teachers with the exception of one had no formal preparation, i.e. in-service training related to teaching low attainers in primary mathematics.

Table 4
Profile of teacher participants of the study

Attribute	Number of teachers
Gender	
Male	5
Female	18
Age (years)	
Less than 30	6
30 – 39	11
40 – 49	6
Teaching primary math (years)	
Less than 1	3
1 – 3	8
4 – 6	1
7 – 10	6
More than 10	5
Teaching math to P3/P4 pupils (years)	
Less than 1	3
1 – 3	15
4 – 6	3
7 – 10	1
More than 10	1
Teaching low attainers (years)	
Less than 1	10
1 – 3	11
More than 4	2
Attended in-service training related to teaching low attainers in primary maths	
Yes	1
No	22

5.2 *Instruments*

In this section we describe briefly all the instruments that were used to collect data for the study. All the instruments were developed by the research team of the study. In chapters 2, 3 and 4 the instruments used and data collected to answer the three research questions respectively will be detailed.

5.2.1 *School management questionnaire*

The School Management Questionnaire was used to gather data from school leaders about the practices used in their schools to identify, organise, and motivate the low attaining pupils, particularly in mathematics. Appendix A shows the content of the questionnaire.

5.2.2 *Teacher profile questionnaire*

The Teacher Profile Questionnaire was used to gather data from the teachers in the study about their qualifications (academic and professional), type of appointment in the teaching profession (teacher, subject head, head of mathematics department, relief teacher, adjunct teacher), their experience in teaching primary mathematics (in terms of number of years), their experience in teaching mathematics to Primary 3 / 4 pupils, their experience teaching low ability pupils (in general), low ability pupils in mathematics, and any specific training related to teaching low attaining pupils. Appendix B shows the teacher profile questionnaire.

5.2.3 *Teacher interview prompts*

As part of the study teacher participants were interviewed. The purpose of the interview was to find out teacher's understanding of "low attainers", and their school's practices related to identification, organisation and motivation of low attainers in mathematics. It also sought information on practices specific to the teacher. The interview was guided by the prompts shown in Appendix C.

5.2.4 *Benchmark mathematics test*

The eight benchmark paper and pencil mathematics tests were used to benchmark the performance of the pupil participants in mathematics. These tests are described in detail in chapter 2 of the book. Appendices D1–D5 show the tests.

5.2.5 *Pupil profile questionnaire*

The Pupil Profile Questionnaire was used to gather data from the pupils about their background such as, race, dwelling, education and occupation of parents, language spoken at home, availability of study desk, time spent on homework, help with homework, out-of-school help with mathematics (tuition), mathematics related activities with parents and family members at home or out of school. Appendix E shows the content of the questionnaire.

5.2.6 *Pupil behaviour questionnaire*

The Pupil Behaviour Questionnaire was completed by the teachers for every pupil in the study they taught. It gathered data on their classroom interactions, on-task behaviours, work attitudes and attendance. Appendix F shows the content of the questionnaire.

5.2.7 *Pupil maths learning questionnaire*

The Pupil Maths Learning Questionnaire was used to gather data from the pupils about their home support, teachers' classroom pedagogy, beliefs about mathematics, self-confidence, effort and learning preferences. Appendix G shows the content of the questionnaire.

5.2.8 *Pupil interview prompts*

The pupils were interviewed in small groups. The size of the group was four or five pupils. During the interview the pupils were prompted by the researchers to talk about their home support, perception of mathematics and self (confidence, effort / motivation, learning preferences) and the classroom pedagogy of their mathematics teachers. Appendix H shows the interview prompts.

5.2.9 *Pupil journal*

The pupils in the study were given a journal book for the school year 2010. They were encouraged to write as often as possible about their feelings, likes and dislikes about learning mathematics in their journals. Appendix I shows a page from the journal book.

5.2.10 *Lesson observation analytical questions*

As part of the study, lessons attended by the pupils and taught by their mathematics teachers were observed. This opportunity for the researchers was strictly voluntary on the part of the teachers in the study. The lesson observation was guided by analytical questions such as the focus of the lesson, instructional sequence, tailoring of instructions to meet the needs of different learners, characteristics of tasks used and the type of learning environment created by the teacher. Appendix J shows the list of the questions.

References

Alderman, K. (1990). Motivation for at-risk students. *Educational Leadership*, *48*(1), 27-30.

Blatchford, P., Kutnick, P., Baines, E. & Galton, M. (2003). Toward a social pedagogy of classroom group work. *International Journal of Educational Research*, 39(1), 153-172.

Boaler, J. (1997). Setting, social class and survival of the quickest. *British Educational Research Journal, 23*(5), 575-595.

Boaler, J., William, D., & Brown, M. (2000). Students' experiences of ability grouping - disaffection, polarisation and the construction of failure. *British Educational Research Journal, 26*(5), 631-648.

Borich, G.D. (1992). *Effective teaching methods* (2nd ed). Upper Saddle River, NJ: Merrill/Prentice Hall.

Bringham, F.J., Scruggs, T.E., & Mastropieri, M.A. (1992). Teaching enthusiasm in learning disabilities classrooms: Effects on learning and behaviour. *Learning Disabilities Research & Practice*, *7*, 68-73.

Christenson, S.L., Ysseldyke, J.E., & Thurlow, M.L. (1989). Critical instructional factors for students with mild handicaps: An integrative review. *Remedial and Special Education*, *10*(5), 21-31.

Cockburn, A.D. (1999). *Teaching mathematics with insight: The identification, diagnosis and remediation of young children's mathematical errors*. London: Falmer Press.

Day, C., Stobart, G., Sammons, P., Kington, A., Gu, Q., Smees, R., & Mujtaba, T. (2006). *Variations in teachers' work, lives and effectiveness (VITAE)*. Nottingham: DfES.

Denvir, B., Stolz, C., & Brown, M. (1982). *Low attainers in mathematics 5-16: Policies and practices in schools*. Methuen Educational.

Dunne, M., Humphreys, S., Sebba, J., Dyson, A., Gallannaugh, F. & Muijs, D. (2007). *Effective teaching and learning for pupils in low attaining groups*. London: Department for children, schools and families.

Foong, P.Y. (1999). Can we identify the reasons for a pupil's low attainment in mathematics. In S.P. Loo (Ed.), *Educational Challenges in the New Millennium* (pp. 648-660). Proceedings of the MERA-ERA joint conference, Malaysia.

Ginsburg, A., Cooke, G., Leinward, S., Noell, J., & Pollock, E. (2005). *Reassessing U.S. international mathematics performance: New findings from the 2003 TIMSS and PISA*. Report prepared by the American Institutes for Research for the U.S. Department of Education.

Hallam, S., & Ireson, J. (2007). Secondary school pupils satisfaction with their ability grouping placements. *British Educational Research Journal, 33*, 27 - 45.

Haylock, D. (1991). *Teaching mathematics to low attainers, 8-12*. Paul Chapman Publishing Ltd.

Houssart, J. (2004). *Low attainers in primary mathematics – The whisperers and the maths fairy*. London: Routledge, Falmer.

Ireson, J., Hallam, S., Hack, S., Clark, H., & Plewis, I. (2002). Ability grouping in English secondary schools: Effects on attainment in English, Mathematics and Science. *Educational Research and Evaluation, 8*(3), 299-318.

Kutnick, P., Blatchford, P., & Baines, E. (2002). Pupil groupings in primary school classrooms: sites for learning and social pedagogy? *British Educational Research Journal, 28*, 189-208.

Kutnick, P., Hodgkinson, S., Sebba, J., Humphreys, G., Galton, M., Steward, S., Blatchford, P., & Baines, E. (2006). *Pupil grouping strategies and practices at key stage 2 and 3: Case studies of 24 schools in England*. London: DfES.

Lee, L. (1999). *Self-esteem of pupils with learning difficulties*. Unpublished M Ed. Dissertation. National Institute of Education, Singapore.

Leong, K.Y.E. (1999). *Parental involvement and its effects on academic performance of students*. Unpublished M Ed. Dissertation. National Institute of Education, Singapore.

Lim, K.K.A. (1999). *A comparison between the learning styles of high-achieving and low-achieving pupils in a primary school*. Unpublished M Ed. Dissertation. National Institute of Education, Singapore.

Loo, C.P., & Fong, H.K. (1996). Variables associated with mathematics achievement of primary five pupils in Singapore schools. *The Mathematics Educator, 1*(1), 1-16.

Mercer, C., & Mercer, A. (2005). *Teaching students with learning problems*. NJ, Upper Saddle River: Merrill Publishing Co.

Ng, T.J. (2003). *Possible factors relating to poor performance of underachievers in primary schools*. Unpublished M Ed. Dissertation. National Institute of Education, Singapore.

Slavin, R.E. (1990). Achievement effects of ability groupings in secondary schools: a best evidence synthesis. *Review of Educational Research*, *60*(3), 471 – 499.

Sprick, R.S. (1985). *Discipline in the secondary classroom: A problem-by-problem survival guide*. West Nyack, NY: The Centre for Applied Research in Education.

Chapter 2

Mathematics Content Knowledge of Low Attainers

KOAY Phong Lee   CHANG Suo Hui

Masura GHANI

This chapter examines the performance of 390 Primary 4 low attainers from Singapore across different strands of primary mathematics. The data collected were subjected to simple descriptive statistical analysis and also error analysis. From the analysis of the data, performance on every item of the tests was studied and common errors made by the pupils were identified. The results revealed valuable information for teachers in the areas of instruction, marking scheme and memory strategies.

1 Introduction

Researchers have used various discrete aspects of numerical proficiency such as strategic counting, retrieval of basic number facts, numeral recognition to predict future success in mathematics (e.g., Jordan, Levine & Huttenlocher, 1995; Geary, 2004). In Singapore, a School Readiness Test (SRT) is used to identify Primary 1 pupils who are at risk of serious reading difficulties and / or mathematics difficulties. Instructional programmes such as Learning Support Programme (LSP) for both English and Mathematics are tailored for these pupils. The Learning Support Programme for Primary 1 and 2 pupils was introduced in all schools in January 2007 and January 2008 respectively. However, the programme is not extended beyond Primary 2. Therefore, pupils who continue to fail in mathematics beyond Primary 2 do not have access to a structured programme for help and therefore are at risk henceforth. In the Low Attainers in Primary Mathematics (LAPM) study, the term "low attainers" refers to the pupils who attain very much less in mathematics when compared to their contemporaries, in this case Primary 4 pupils.

Singapore mathematics curriculum is spiral in structure. A mathematic idea originates at the centre of the spiral and is revisited and extended in the following years. Each year, related ideas are introduced. Hence, pupils who are low attainers in mathematics in the preceding years would continue to struggle with mathematics if appropriate help is not rendered to them. Misconceptions they have will be brought forward when they learn the related ideas.

This chapter addresses the first research question of the study: "How did the low attaining pupils perform in mathematics tests that tested mastery of Primary 3 mathematics content knowledge and skills?" It reports on the performance of the low attainers in mathematics in all strands of mathematics found in the Primary 3 mathematics syllabus: (a) whole numbers; (b) fractions; (c) geometry; (d) measurement and (e) data analysis.

2 Background

2.1 *Whole numbers*

Children with mathematics difficulties have been reported to have incomplete mastery of basic number facts, incompetent arithmetic computations, difficulty transferring knowledge, making connections, incomplete understanding of the language of mathematics, particularly in interpreting word problem sentence construction (Jordan, Levine, & Huttenlocher, 1995; Gracia, Jimenez, & Hess, 2006).

In primary school, pupils start their journey of mathematics learning with arithmetic. By the end of Primary 3, pupils are expected to master the basic concepts and skills involving 4-digit numbers (See Appendix D1).

Bryant, Bryant, and Hammill (2000) reported that the top seven most common mathematics difficulties as ranked by a group of special education professionals are as follow:

1. difficulty with word problems
2. difficulty with multi-step problems
3. difficulty with the language of mathematics
4. failure to verify answers and settle for the first answer
5. lack of automaticity with number facts
6. take a long time to complete calculation
7. make "borrowing" (i.e. regrouping, renaming) errors

Solving word problems and computation represent two distinct domains of mathematical cognition in low attainers in mathematics (Fuchs, Fuchs, Hamlett, Stuebing, & Fletcher, 2008). To be able to compute correctly does not necessary lead to success in problem solving.

2.2 *Fractions*

The concept of fraction as part of a whole is introduced in Primary 2. By the end of Primary 3, pupils are expected to master equivalent fractions and addition and subtraction of two related fractions within one whole. Fraction is a difficult concept for lower primary pupils. It is different from whole numbers as it uses two numerals to represent a single quantity. Both the numerals (the numerator and the denominator) affect the magnitude of fractions and by simply looking at the fraction labels, pupils would not be able to identify the larger / smaller fraction. Moreover, unlike whole numbers, different fraction labels can represent the same quantity, for example, $\frac{1}{4}$, $\frac{2}{8}$ and $\frac{3}{12}$ are equivalent. This understanding of equivalent fractions is crucial in the domain of fractions. Primary 3 pupils would not be able to add / subtract related fractions without the knowledge of equivalent fractions.

From the conceptual change perspective (Stafylidou & Vosniadou, 2004), fraction would be a difficult concept for the primary school pupils to master as learning the concept requires the "reorganisation of existing knowledge structures" of the pupils. Pupils who attempt to assimilate the newly taught concept of fraction to their existing knowledge base (of natural numbers) and without making radical reorganisation of their existing knowledge base would often have misconceptions of fraction. For example, pupils who try to apply what they know about the comparison of natural numbers to compare $\frac{2}{3}$ and $\frac{3}{5}$ would conclude that $\frac{3}{5}$ is greater because $3 > 2$ and $5 > 3$.

2.3 *Measurement*

In Primary 1, pupils learn to measure and compare lengths and masses with non-standard units using appropriate terms. Capacity and standard units are introduced in Primary 2. Compound units are introduced in Primary 3, but pupils are not expected to operate on compound units.

They begin to learn to tell and write time to the hour / half hour in Primary 1, to 5 minutes in Primary 2 and progress to 1 minute in Primary 3. By then, they are expected to solve word problems involving duration of a time interval.

Primary 1 pupils also learn to tell the amount of money and solve word problems, in cents and in dollars but not in the combinations of dollars and cents. These concepts and skills are extended in Primary 2 when the decimal notation is introduced. By end of Primary 3, pupils are expected to solve word problems involving addition and subtraction of money in dollars and cents expressed in decimal notation.

Area and perimeter are dealt with in Primary 3. Pupils learn to measure area in cm^2 and m^2. By end of Primary 3, they are expected to apply the area / perimeter formulas to solve word problems involving area / perimeter of squares and rectangles.

2.4 *Geometry*

The learning of geometry begins with the identifying and describing of basic shapes (rectangle, square, circle and triangle) in Primary 1. By end of Primary 3, pupils are expected to identify and name perpendicular and parallel lines and draw these lines on square grids. They are introduced to the concept of angle as an amount of turning, and are expected to identify right angles and compare angles with respect to the right angle. Bishop (1986) identifies the following three potential sources of difficulty for the learners in learning geometry:

- a restricted notion of what is important,
- a narrow view of the representation, and
- the confusion of form with content.

He argued the needs for the learners to mathematise their spatial experiences. They should focus their attention on as many relevant features of space as possible instead of learning the definitions. Hence, teachers should provide a wider approach to representations of geometrical ideas to facilitate learning and avoid misconceptions as shown in Figure 1.

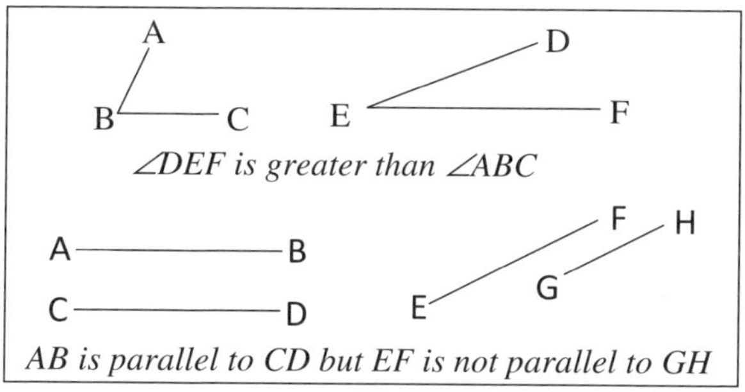

Figure 1. Misconceptions in geometry

Besides drawing, language is also another mode of representing geometrical ideas. Bishop (1986) cautioned against the use of labels prematurely, and advocated the approach of using appropriate activities that emphasise geometrical features, properties and phenomena to give the content of geometry an objective form.

2.5 *Data analysis*

According to Lowrie and Diezmann (2007), there are six graphical languages categorised according to how the information is encoded visually. Picture graphs and bar graphs are examples of the opposed-position language where the information is encoded by a marked set that is placed between two axes in the graph. In the primary school, picture graph without scale is introduced in Primary 1 while picture graph with scale is introduced in Primary 2. By end of Primary 3, pupils are expected to interpret bar graphs involving scales. There are three levels of graph comprehension: reading the data, reading between the data and reading beyond the data (Friel, Curcio, & Bright, 2001). Graph comprehension is not just reading the graph; pupils need to think about the relations represented in the graph, and how the data change over time or across the categories. When the graph involves scale, counting and adding are not sufficient, understanding of multiplicative reasoning is necessary.

3 The Subjects

Nine government primary schools across Singapore participated in the study. A total of 560 Primary 4 pupils were first identified by their teachers as low attainers in mathematics. These are pupils who were at risk of failing their Primary School Leaving Examination in Primary 6. Two of the pupils were later excluded from the study when it was found that they had serious learning disabilities. They could neither write their names properly nor answer any test items correctly. Some pupils were absent during some of the testing sessions, hence the number of pupils who completed the tests varied from test to test. Only 390 pupils completed all the tests and their data are presented in this chapter.

4 Procedures and Instruments

Eight written tests were used to measure the pupils' proficiency level on Primary 3 mathematics content (see Table 1). The benchmark tests were administered to the pupils, one at a time, by their teachers during instructional time. No two tests were administered on the same day. As the tests assessed mathematical knowledge that the pupils were expected to know, the pupils were not informed of the test in advance and hence did not make special preparation for the tests.

Table 1
Topics assessed in the benchmark tests

Test	Strands
1	Whole numbers (Part I) ▪ reading, writing, interpreting 4-digit numbers ▪ comparing and ordering 4-digit numbers ▪ completing number sequences
2	Whole numbers (Part II) ▪ understanding the four operations in context ▪ carrying out addition & subtraction algorithms ▪ recalling multiplication facts
3	Whole numbers (Part III) ▪ solving up to 2-step word problems
4	Fractions ▪ reading, writing and representing simple fractions ▪ comparing and ordering fractions ▪ simplifying fractions ▪ adding & subtracting related fractions ▪ solving 1-step word problem
5	Measurement (Part I) ▪ comparing lengths ▪ measuring lengths using rulers ▪ converting measures in lengths, capacity, mass and money ▪ finding area and perimeter of composite figures draw on grid
6	Measurement (Part II) ▪ solving up to 2-step word problems
7	Geometry ▪ identifying straight and curve lines, parallel and perpendicular lines ▪ identifying shapes ▪ identifying and comparing angles ▪ completing patterns ▪ identifying cubes and cuboids
8	Data analysis ▪ reading data in picture graph and bar graph ▪ reading between data in picture graph and bar graph ▪ completing a bar graph

5 Data and Findings

The mean and standard deviation of the scores for each test are shown in Table 2. Pupils seem to perform relatively well in whole number operations and consistently poorly in word problem tests.

Table 2
The maximum score, mean and standard deviation of the tests

Test	Maximum Score	Mean	Standard deviation
Whole numbers (Part I)	30	17.10	4.49
Whole numbers (Part II)	29	22.10	3.98
Whole numbers (Part III)*	9	3.52	2.45
Fractions	29	13.71	6.98
Measurement (Part I)	21	10.14	3.69
Measurement (Part II)*	8	3.20	2.05
Geometry	10	6.34	2.16
Data analysis	10	6.37	2.55

* word problems test

The results of pupil performance on test items will be discussed by strands. Only responses with $n > 7$ (2% of 390) are reported. Common learning difficulties are highlighted. The asterisk indicates the correct answer.

5.1 *Whole numbers*

Whole numbers were tested in three parts. Part I, comprising Items 1 to 17, tested basic concepts of whole numbers. Part II, comprising Items 18 to 22, tested the four operations on whole numbers. Part III, comprising Items 23 to 31, tested one-step and two-step word problems involving whole numbers.

5.1.1 Number notation

The representation of numbers comes in many forms such as numerals, written word in English, base-ten system (Hindu-Arabic system) and number line. Pupils had little difficulty reading and writing numbers (see Table 3a, Items 1 and 2), but found it difficult to interpret the numbers in terms of place value (see Table 3a, Item 5), or representing the numbers on number lines (see Table 3b, Item 17).

From Table 3a, comparison between Items 1 and 2 revealed that pupils were more proficient in the numeral than the written English word form of numbers. The problem stemmed from a poor command of English Language particularly spelling. This is supported by similar problems with the spelling of words like "hundred" and "thousand". This is not surprising considering that these pupils struggled with English Language scoring an average of 47.01% for their Primary 3 Semestral Examination 1. In fact, this poor command of the language puts them at a disadvantage in solving word problems which will be dealt with later.

Items 4 and 5 covered the concept of base-ten system. Pupils performed poorly with only 63% and 42% of them were able to do Items 4 and 5 respectively. Base-ten system is a highly sophisticated and powerful system that enables large numbers to be expressed without the need to reinvent more symbols. However, differentiation between the base-ten notation and the English word form of a number is seldom made explicitly to the pupils. As a result, pupils were lost in the process of understanding the various notations and were overwhelmed by the cognitive demand of converting notations. Moreover, the English Language, unlike other Asian languages such as Chinese, does not support the conceptual structure of multi-digit base-ten system. For example, the English word "thirteen" bore no resemblance to its corresponding base-ten notation of "one ten, three ones" (Geary, 1996). More effort and extra steps were needed to connect the two different notations. As such, despite recognising the place value at which the digit was stated (e.g. Item 4a, digit 8 is in the tens place and stands for 8 tens or eighty), pupils failed to present the correct response in accordance to the requirements of base-ten system. Thus their responses had either omitted the letter "*s*" or failed to state the value. Such conventions, according to Geary (1996), must be taught explicitly through direct teaching.

Number line is a commonly used instructional tool in primary school for the development of number concepts. It is also part of the everyday life such as thermometers. Table 3b, shows a stark difference in the percentages of correct response between Items 17a and 17b revealing the misconceptions and difficulties faced by the pupils in representing numbers on the number line. Both items fell under the category of structured number line where mathematical information is represented by the placement of marks in proportional segments on a horizontal line. According to Diezmann and Lowrie (2006), the successful interpretation of number line among pupils exhibited an understanding of the tool as a measurement model rather than a counting model. They employed strategies related to specific distance, the proximity of the numbers or the reference points and paid attention to the distance between the marks. In addition, the unit of iteration varies depending on the reference points and the number of markings along the line. Petitto (1990) suggested that younger children perceived numbers on the line as a sequence of numbers ordered in relation to their magnitude. But older children perceived the numerical quantities primarily in terms of proportion which aid them in visualising midpoints or references as landmarks.

The proportional concept is an important aspect to the appreciation and understanding of units of iterations. As in Item 17a, the unit of iteration is one whereas that of Item 17b is ten. 61.54% of the pupils mistook the unit of iteration in Item 17b to be the same as Item 17a and reported the answer as 2782 and 2785. The mistake could have arisen from two possible reasons. First, pupils might have counted the ten segments between the two given numbers and assumed that the unit of iteration is one. They had ignored the two reference points. Second, pupils could be careless and presumptive. Unfortunately, the result is inconclusive until further investigation but the stark difference in the percentages between the two items signifies a deep problem with unit of iteration of the measurement model of number line.

Table 3a

Items on Number notation — Distribution of common responses

Item	Response	n (%)
1. Write 5802 in words.	Five thousand, eight hundred and two*	339 (73.85)
2. Write eight hundred and thirty two in numerals.	8032*	346 (88.72)
	832	9 (2.31)
	80032	10 (2.56)
4. In 9853, the 8 stands for 8 hundreds. a) In 9083, the 8 stands for _______	8 tens/eighty*	247 (63.33)
	Tens	88 (22.56)
	8 ten	13 (3.33)
	Ten	10 (2.56)
b) In 8930, the 8 stands for ___________	8 thousands/8000*	163 (41.79)
	8 thousand	92 (23.59)
	thousand	52 (13.33)
	thousands	40 (10.26)
5a) In 8906, the 9 is in the ________ place.	hundreds* *(including error in spelling)*[1]	161 (41.28)
	hundred	136 (34.87)
	(including error in spelling)[2]	16 (4.10)
	9 hundreds	10 (2.56)
	9 hundred	30 (7.69)
	900	37 (9.49)
b) In 7004, the 7 is in the ________ place.	thousands*	134 (34.36)
	thousand	165 (42.31)
	(including error in spelling)[3]	12 (3.08)
	7 thousands	10 (2.56)
	7 thousand	32 (8.21)
	7000	37 (9.49)

Note: [1,2]*Common forms of misspelled "hundred" — hunder, hunderd, hurndred, hundered*
[3]*Common forms of misspelled "thousand" — thousad, thosand, thoundsand, tonsand*

Table 3b

Items on Number notation — Distribution of common responses

Item	Response	n (%)
17. Fill in the missing numbers in the blanks.	3157, 3164*	302 (77.44)
	with 3164 correct	21 (5.38)
a) 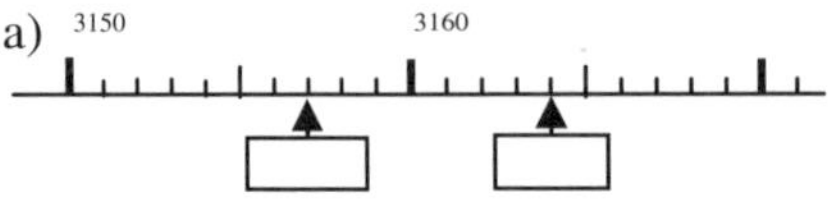		
b) 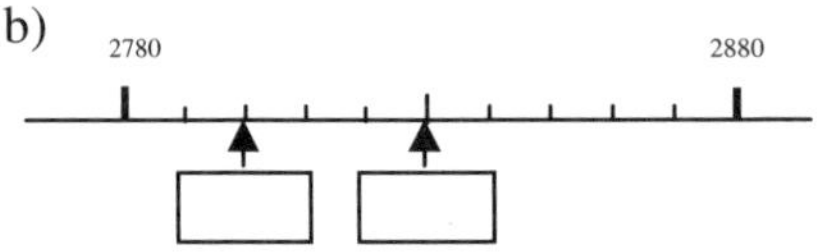	2800, 2830*	26 (6.67)
	with 2800 correct	11 (2.82)
	2782, 2785	240 (61.54)
	No response	7 (1.79)

5.1.2 Ordinal numbers

The key perceptual cue to the order of the animals, Item 3 was the food stall (Table 4). As seen in Item 3a, 83.85% of the pupils were able to pick up the cue while only 10.77% of them assumed the left to right rule, ignoring important perceptual and contextual cue i.e. the stall and the direction to which the animals are facing. Language played an important role in Items 3b and 3c. The phrase "comes just before" can be complex and posed a problem to these pupils. Pupils who chose the response "bird" could have been misled by the word "before". Similarly, in Item 3c, a total of 298 (260 + 8 + 30) of them could identify that the bird is in the second position but only 260 could decode it correctly in English. Thus these pupils did not appear to have problems with the conceptual understanding of ordinal numbers but rather a more pertinent language problem.

Table 4
Items on Ordinal numbers — Distribution of common responses

Item	Response	n (%)
3. Look at the picture and answer the following items. The animals are queuing up for food.  dog rabbit hen bird turtle		
a) The animal at the end of the queue is the _______.	dog*	327 (83.85)
	turtle	42 (10.77)
b) The hen comes just before the ________ .	rabbit*	204 (52.31)
	bird	167 (42.82)
c) The bird is the _________ animal in the queue.	2^{nd*}	260 (66.67)
	2st. 2sd, 2th	8 (2.05)
	2	30 (7.69)
	$4/4^{th}$	19 (4.87)
	hen	15 (3.85)
	turtle	11 (2.82)

5.1.3 Comparing numbers

The concept of comparing numbers were examined through four different item types namely; (I) Items 6, 7 and 10, (II) Items 8 and 9, (III) Items 11 and 16 and (IV) Items 12 and 13.

Items 6, 7 and 10 in Table 5 test pupils' ability to count up or down by 1, 100 and 1000. The pupils seemed very proficient in such items but suspicion was raised that the underlying strategy adopted by these pupils might be in contrary to the intention of such items. The objective delineated for such items was the development of pattern looking and understanding relationships of more / less than which lay the foundation to number sequence and algebraic thinking. The desired outward expression of the mastery of the appropriate mathematical thinking will be evidence of use of a number line or the common practice of using an arrow ( ) to signify some form of pattern looking. However, as shown in the pupils sample in Figure 2, they were found using the vertical algorithm. This is not surprising considering pupils' weak performance with the use of number line in Item 17.

Table 5
Items on Comparing numbers: Type I — Distribution of common responses

Item	Response	n(%)
6a) 879 —Add 100→ ☐	979*	367 (94.10)
b) 4032 —Add 100→ ☐	4132*	361 (92.56)
	5032	10 (2.56)
7a) 5249 —Subtract1000→ ☐	4249*	343 (87.95)
	6249	19 (4.87)
b) 1876 —Subtract1000→ ☐	876*	344 (88.21)
	2876	17 (4.36)
10. Write a number that is one more than 9899.	9900*	200 (51.28)
	9999	86 (22.05)
	10 000	30 (7.69)
	1000	10 (2.56)

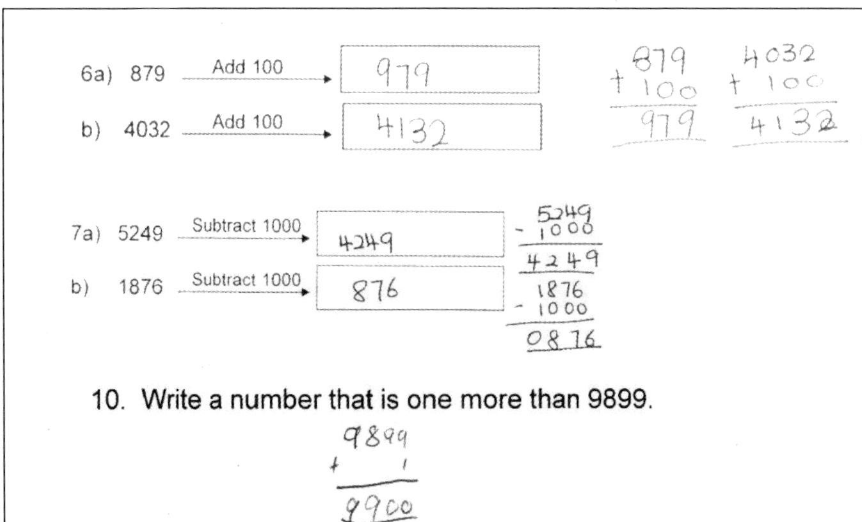

Figure 2. Sample of pupils' answers to Type I items

Table 6 shows the pupils' answers to Items 8 and 9. From the responses of "459", "8301" and "8031", pupils were observed to experience difficulties dealing with zeros which is in congruent to vertical algorithm of addition and subtraction discussed later. The function of zero in the base-ten system as a placeholder is in contrary to the understanding of zero as "nothing" in daily lives resulting in pupils' tendency to omit it as in "459" or a convenient deposit of it disregarding its significance. Moreover, the idea and distinction of digits and numbers is seldom covered in school resulting in an incomplete understanding of number system. The problem is made worse with a practice of avoiding items of such nature involving zeros.

Table 6

Items on Comparing numbers: Type II — Distribution of common responses

Item	Response	n (%)
8. Form the smallest **4-digit** number with these digits.	4059*	261 (66.92)
0 4 5 9	459	52 (13.33)
	4509	20 (5.13)
	4590	26 (6.67)
9. Form the largest **4-digit** number with these digits.	8310*	285 (73.08)
1 0 8 3	8301	29 (7.44)
	8031	18 (4.62)
	138	9 (2.31)

As shown in Table 7, least number of pupils answered Items 11 and 16b correctly among the series of items on comparing numbers. Plausible reasons might be language problems or the lethal combinations of straining the working memory capacity and the need to activate the number line as a form of linear visual representation.

In the former, responses such as "5167", "5367", "5467", "100" and "10634" reflected a poor comprehension of the task demands. Attention was paid only to single key words like "greater" that resulted in "5367" and "5467" or "less than" resulting in "5167". Alternatively, pupils might add (resulting in 10634) or subtract (resulting in 100) the two given numbers depending on which word they focused on; "greater" as more or "less than" as subtract.

To answer Item 11 correctly, pupils were required to hold two key information, "greater than 5267" and "less than 5367", and simultaneously depict them on a mentally constructed number line in the working memory. After which they were required to retrieve all the consecutive terms in that imaginary number line before they picked one number that fulfil the requirement of the item. Geary (2004) commented that children with mathematics learning difficulties may suffer from a deficit in information representation and manipulation, visuospatial competencies as well as that in the central executive control over attention, all of which were evident processes in Item 11.

Table 7

Items on Comparing numbers: Type III — Distribution of common responses

Item	Response	n (%)
11. Write a number that is greater than 5267 but less than 5367.	5268≤ n ≤5366*	146 (37.44)
	5167	35 (8.97)
	5367	12 (3.08)
	5467	12 (3.08)
	10 634	15 (3.85)
	100	89 (22.82)
16. Circle **one** number that has a value nearest to the number on the left. a) 644 → 650 600 700 640	640*	244 (62.56)
	600	50 (12.82)
	650	45 (11.54)
	700	9 (2.31)
	No response	22 (5.64)
b) 2097 → 2000 2100 2090 2095	2095*	148 (37.95)
	2100	125 (32.05)
	2090	47 (12.05)
	2000	23 (5.90)
	No response	25 (6.41)

As shown in Table 8, pupils performed relatively well in Item 12 (about 78% correct) and Item 13 (about 87% correct) signifying pupils' proficiency in such tasks. Unlike the items mentioned above, the numbers were laid out before them thus alleviating demands on the working memory to hold and retain the information. The technique adopted most likely was the comparison of digits in respective place value systematically one after the other. However, proficiency in this item might stem from a mastery of the procedure rather than a genuine understanding of the value of respective numbers posed.

Table 8
Items on Comparing numbers: Type IV — Distribution of common responses

Item	Response	n (%)
12. Arrange these numbers from the smallest to the largest. 6402 6240 4062 4602	4062, 4602, 6240, 6402*	304 (77.95)
	4062, 4602, 6402, 6240	27 (6.92)
	4602, 4062, 6402, 6240	14 (3.59)
	6402, 6240, 4602, 4062	11 (2.82)
13. Arrange these numbers from the largest to the smallest. 398 908 389 983	983, 908, 398, 389*	339 (86.92)
	983, 908, 389, 398	12 (3.08)
	389, 398, 908, 983	11 (2.82)

5.1.4 Even numbers

The topic on odd and even numbers comprises of a very small portion of the number concepts to be mastered in Primary 3. As seen in Table 9, the responses from the pupils were randomised with no apparent pattern in their error. Four possible reasons are proposed:

- a lack of emphasis on the concept of odd and even in the syllabus. Upon examination of the two common textbook series in Singapore, only one included two-page coverage on this concept through divisibility whereas the other has none.
- the numbers catered in the textbook were not more than 150.
- the concept is introduced together with division with remainder and unless explicit teaching is conducted to highlight this special group of numbers, pupils may not fully appreciate the concept and merely dismiss it off as dealing with another remainder.
- pupils tend to perceive even numbers as those ending with 0, 2, 4, 6 and 8 rather than from a divisibility principle.

Table 9
Items on Even numbers — Distribution of common responses

Item	Response	n (%)
14. Circle all the even numbers.	230, 718*	95 (24.36)
230 189 457 718	230, 189/457	57 (14.62)
	718, 189/457	27 (6.92)
	230, 718, 189/457	19 (4.87)
	230	42 (10.77)
	718	33 (18.46)
	Wrote all the numbers	28 (7.18)
	No response	8 (2.05)

5.1.5 Number patterns

Table 10 shows that pupils were more proficient in completing an ascending pattern (Item 15a) than descending pattern (Item 15c). They were least successful in item involving multiplicative pattern (Item 15b).

The results in Table 10 also show that pupils were more apt at identifying the missing terms than explaining the rule. Mathematical communication, introduced in the 2007 Singapore Mathematics syllabus (Ministry of Education, 2006), remained lacking and weak among these pupils. From the incorrect responses of pupils for Item 15b, it was observed that pupils tend to favour addition to any other operations especially multiplication. This same observation is made in solving word problems where their first choice of operation in most scenarios was addition.

Table 10

Items on Number patterns — Distribution of common responses

	Item	Response	n (%)
15	Find the missing numbers in the number patterns. a) 302, 297, 292, 287, _____	282*	246 (63.08)
		292	19 (4.87)
		272	12 (3.08)
		277	10 (2.56)
	ai) What is the rule? Rule:_____________________	Minus 5*	162 (41.54)
		Add 5	48 (12.31)
		Add 2	21 (5.38)
		No response	72 (18.46)
	b) 2, 4, 8, ________ , 32	16*	94 (24.10)
		11	14 (3.59)
		10	47 (12.05)
		12	54 (13.85)
		14	31 (7.95)
		18	9 (2.31)
		20	11 (2.82)
		22	12 (3.08)
		24	37 (9.49)
		30	15 (3.85)
		No response	13 (3.33)
	bi) Rule:_____________________	Multiply 2*	47 (12.05)
		Add 2	40 (10.26)
		Add 4	34 (8.72)
		Add 6	18 (4.62)
		Add 8	32 (8.21)
		Add 10	10 (2.56)
		Add 12	14 (3.59)
		Add 24	11 (2.82)
		No response	86 (22.05)
	c) 78, _____ , 278, 378, ____	178, 478*	280 (71.79)
		No response	10 (2.56)
	ci) Rule:_____________________	Add 100*	207 (53.08)
		Add 78	10 (2.56)
		Add 1	18 (4.61)
		No response	84 (21.54)

5.1.6 Column addition and subtraction

As seen from the results in Tables 11a and 11b, at least 90% of the pupils were able to execute the vertical algorithm for both addition and subtraction with or without renaming proficiently. As shown by Item 21e (see Table 11b) only subtraction with renaming involving zeros in the minuend has a lower percentage of correct responses (less than 70%). As mentioned before (see Table 6, Items 8 and 9), pupils have difficulties handling zeros. In both cases of a single zero in the tens place as well as two zeros in ones and tens place, pupils were seen putting down a small digit "1" at the left top corner but failed to reduce it to "9". This stems from an incomplete or misunderstood knowledge of the principle of exchange (i.e. trading). Pupils were carrying out the procedure without understanding the rationale. Moreover, when renaming, the digit 1 at the left top corner have two different meanings in addition and subtraction; one more of the larger place value unit in addition renaming but ten of the smaller place value unit in subtraction renaming.

Table 11a

Items on Column addition and subtraction — Distribution of common responses

Item		Response	n (%)
Addition without renaming			
20a)	1 2 7 + 4 2	169*	372 (95.38)
20b)	4 2 3 + 2 6 1	684*	377 (96.67)
Addition (renaming once)			
20c)	2 6 4 + 1 2 6	390*	374 (95.90)
Addition (renaming twice)			
20d)	3 7 8 + 1 4 5	523* 513	351 (90.00) 8 (2.05)
20e)	2 4 7 + 1 5 3	400*	372 (95.38)
Addition (renaming thrice)			
20f)	6 0 7 + 3 9 3	1000* 1090	350 (89.74) 13 (3.33)

Table 11b

Items on Column addition and subtraction — Distribution of common responses

Item		Response	n (%)
Subtraction (without renaming)			
21a)	286 − 25	261*	370 (94.87)
21b)	647 − 326	261*	370 (94.87)
Subtraction (renaming tens)			
21c)	739 − 497	242*	314 (80.51)
		362	23 (5.90)
Subtraction (renaming twice)			
21d)	456 − 278	178*	307 (78.72)
		188	8 (2.05)
Subtraction (zero at tens place of minuend)			
21e)	806 − 439	367*	269 (68.97)
		377	16 (4.10)
		467	13 (3.33)
		477	8 (2.05)
Subtraction (zero at ones and tens place of minuend)			
21f)	600 − 437	163*	291 (74.62)
		173	13 (3.33)
		237	14 (3.59)
		263	17 (4.36)

5.1.7 Multiplication facts

Pupils were well-versed with the multiplication facts. More than 80% of the pupils were able to retrieve the multiplication facts (see Table 12).

Table 12
Items on Multiplication facts — Distribution of common responses

Item	Response	n (%)
Finding the products.		
22. Fill in the missing numbers		
a) 4 × 6 = _____	24*	361 (92.56)
b) 5 × 8 = _____	40*	373 (95.64)
c) 9 × 7 = _____	63*	312 (80.00)
	72	14 (3.59)
d) 6 × 8 = _____	48*	314 (80.51)
	56	8 (2.05)
Finding the missing factors.		
e) _____ × 7 = 14	2*	365 (93.59)
f) _____ × 4 = 28	7*	336 (86.15)
	6	11 (2.82)
g) 5 × _____ = 30	6*	367 (94.10)
h) 7 × _____ = 56	8*	312 (80.00)
	6	13 (3.33)
	7	24 (6.15)

5.1.8 Quotient and remainder

The results in Table 13 show that pupils appeared more familiar with the word "remainder" (68% correct responses) than with "quotient" (57% correct responses). But ironically, there were pupils who could arrive at the remainder yet not obtained the correct quotient (difference of 10%). The long division algorithm is noted as one of the most difficult procedure to master among pupils. It involves multiple steps incorporating all four operations within that same column and is the only algorithm that starts from the left, contrary to other algorithms.

Table 13

Items on Quotient and remainder — Distribution of common responses

Item	Response	n (%)
19. Divide 122 by 8.	15*	224 (57.44)
What is the quotient?	2	8 (2.05)
What is the remainder?	2*	267 (68.46)
	0	29 (7.44)
	1	14 (3.59)
	4	9 (2.31)
	6	12 (3.08)

5.1.9 Identification of correct operation in a given context

As shown in Table 14, Items 18a to 18g composed of different structures of addition, subtraction, multiplication and division in varying context involving money, measurement and sets. An overview of the problem structures and accompanying item is depicted in Figures 3 and 4.

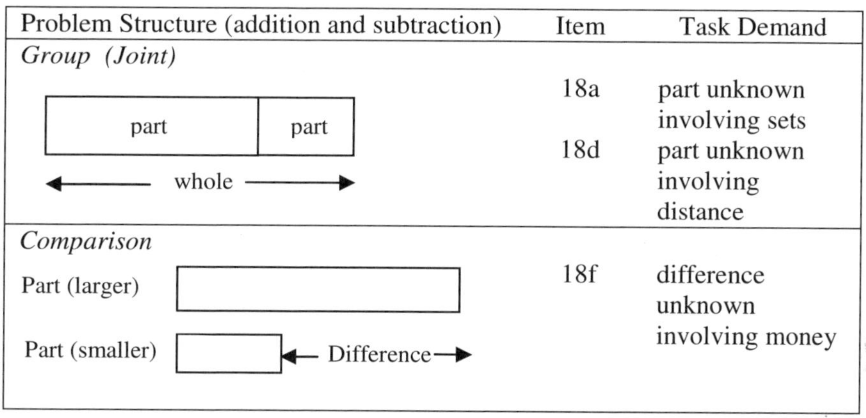

Figure 3. Overview of problem structure (addition and subtraction)

It is apparent from Table 14 that 58.97% of the pupils recognised the structure of Item 18a but only 36.67% and 40.77% of them respectively recognised Items 18d and 18f as addition and subtraction problem structure. The weaker than expected performance achieved could be a result of some learning issues with word problem namely: semantic structure for direct modelling (Carpenter, Moser, & Bebout, 1988), the artificiality or a lack in immediacy of measurement context especially involving length and distance (Haylock, 1991), the primitive intuition model (Fischbein, Deri, Nello, & Marino, 1985) and language problems (Newman, 1983).

The semantic structure of Item 18d is more direct than Item 18a but yielded poorer results because of the context of distance. Whereas for Item 18a, the phrase "bought another" could have misled pupils to activate the key-word strategy to add rather than to comprehend the whole problem of finding the missing part. Similarly for Item 18f, the key word "more" might lead pupils to add. The tendency for pupils to response in a stimulus-response fashion to key words rather than to the logical structure of the problem is detrimental to pupils' problem solving ability. In addition, the comparison structure of Item 18f is deemed most challenging and complex in comparison to the Joint and Change structures.

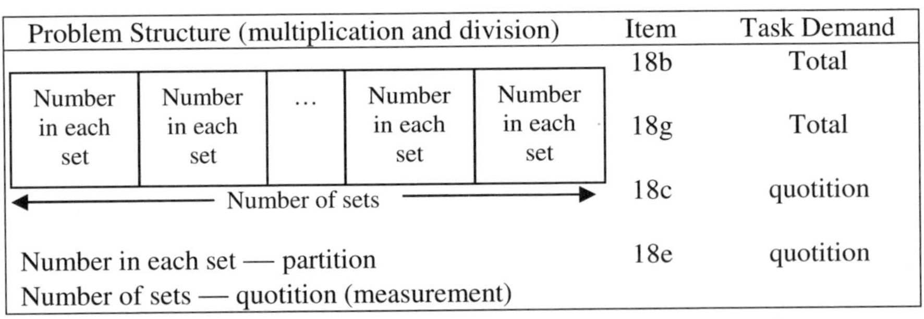

Figure 4. Overview of problem structure (multiplication and division)

Among the items involving multiplication and division, Item 18b has the highest number of correct responses (71%) because its semantic structure facilitates both direct modelling and primitive intuition model meaning. The structure of the problem and the sequence in which information is laid out (cost of one apple and then the cost of five apples) support the primitive interpretation of multiplication as repeated addition where the multiplier and multiplicand are small numbers. Although Item 18g has a similar structure as Item 18b facilitating direct modelling, the large multiplier "30" could mislead pupils to divide instead given about 24% of them chose "30 ÷ 6".

About 48% and 52% of the pupils respectively answered Items 18c and 18e correctly. This is in accordance to other research (Fischbein, et. al., 1985; Yeap & Kaur, 2001) where quotition model is contrary to the primitive intuition model and needs acquisition through explicit instruction thus more demanding to pupils.

Table 14
Items on Identification of correct operation in a given context — Distribution of common responses

Item	Response	n (%)
18. Circle the operation you will use.		
a) Ali had some guppies.	280 – 10*	230 (58.97)
He bought another 10 guppies.	280 + 10	109 (27.95)
He has 280 guppies now.	280 × 10	28 (7.18)
How many guppies did he have at first?	280 ÷ 10	17 (4.36)
b) An apple costs 45¢.	45 × 5*	276 (70.77)
What should the price of 5 apples be?	45 + 5	47 (12.05)
	45 – 5	17 (4.36)
	45 ÷ 5	42 (10.77)
c) A bar of chocolate can be broken into	18 ÷ 3*	189 (48.46)
18 similar squares.	18 + 3	42 (10.77)
There are 3 squares in a row.	18 – 3	20 (5.13)
How many rows are there in the	18 × 3	134 (34.36)
chocolate bar?		
d) Mr Lim walks 12 km home from work.	12 – 4*	143 (36.67)
He stops at a shop after walking 4 km.	12 + 4	96 (24.62)
How much further has Mr Lim to go?	12 × 4	99 (25.38)
	12 ÷ 4	46 (11.79)
e) Salma packed 36 cookies in some bags.	36 ÷ 4*	204 (52.31)
She put 4 cookies in each bag.	36 + 4	47 (12.05)
How many bags did she have?	36 – 4	29 (7.44)
	36 × 4	101 (25.90)
f) Ali has $45 and Meng Ho has $5.	45 – 5*	159 (40.77)
How much more does Ali have than	45 + 5	113 (28.97)
Meng Ho?	45 × 5	58 (14.87)
	45 ÷ 5	49 (12.56)
g) An oven tray can hold 6 cakes.	30 × 6*	209 (53.59)
Mrs Tan fills 30 trays.	30 + 6	57 (14.63)
How many cakes will Mrs Tan bake?	30 – 6	24 (6.15)
	30 ÷ 6	93 (23.85)

5.1.10 Word problems

Tables 15a and 15b show pupils' performance in solving word problems.

Table 15a
Items on Word problems — Distribution of common responses

Item	Response	n (%)
One-step (involving division)		
23. 30 cards are shared out equally among 5 children. How many cards does each child have?	$30 \div 5 = 6$*	279 (71.54)
	$30 + 5 = 35$	17 (4.36)
	$30 \times 5 = 150$	31 (7.95)
	$30 - 5 = 25$	19 (4.87)
25. Each ticket costs $6. How many tickets can be bought with $48?	$48 \div 6 = 8$*	176 (45.13)
	$48 \div 6 \neq 7$	14 (3.59)
	$48 - 6 = 42$	19 (4.87)
	$48 + 6 = 54$	26 (6.67)
	$48 \times 6 = 288$	78 (20.00)
Two-steps (involving addition and subtraction)		
24. Jane has $100. She earns another $50 more and spends $80. How much does she have now?	$100 + 50 - 80 = 70$*	244 (62.56)
	$100 - 80 = 20$	8 (2.05)
	$80 + 50 - 100 = 30$	27 (6.92)
	$80 + 50 = 130$	12 (3.08)
26. I buy two books costing $14 and $20. What is my change from $50?	$50 - (14 + 20) = 16$*	229 (58.71)
	$50 - (14 + 16) \neq 24$ *(failed to rename)*	19 (4.87)
	$20 - 14 = 6$	16 (4.10)
	$14 + 20 = 34$	16 (4.10)
	$50 + 14 - 20 = 44$	8 (2.05)
	$50 - 14 + 20 = 56$	11 (2.82)
	$50 + 14 + 20 = 84$	26 (6.67)
30. Mr Ali had 5000 bottles of cooking oil in his store room. He sold 895 bottles in the first month and 1042 bottles in the second month. How many bottles of cooking oil had he left?	$5000 - 895 - 1042 = 3063$*	145 (37.18)

Table 15b
Items on Word problems — Distribution of common responses

Item	Response	n (%)
Two steps (involving multiplication or division and addition or subtraction)		
27. Siti packs some tarts in 9 boxes. She packs 4 fruit tarts and 5 egg tarts in each box. How many tarts does she pack altogether?	(4 + 5) × 9 = 81* No response	114 (29.23) 8 (2.05)
28. Sally wants to buy a piano that costs $3010. She saves $200 each week for 3 weeks. How much more money does she need to save?	3010 – (200 × 3) = 241* No response	76 (19.49) 8 (2.05)
29. Rohani has 50 envelopes and 220 cards. She puts 5 cards in each envelope. How many envelopes are left unused?	50 – (220 ÷ 5) = 6* No response	26 (6.67) 18 (4.62)
Two steps (comparison, involving addition)		
31. 400 concert tickets were sold on Saturday. 163 more concert tickets were sold on Sunday than on Saturday. How many concert tickets were sold on both days?	163 + 400 + 400 or 163 + (2 × 400) = 963* No response	83 (21.28) 26 (6.67)

The poor performance of the pupils in solving word problems can be discussed in the light of three areas; the semantic structure of the problem, computational proficiency of the pupils and choice of operation by pupils. A breakdown of the problem semantic structures and relations using Yeap and Kaur (2001) and the number of pupils committing computational error are shown in Table 16.

Table 16
Semantic relations and computational errors made

Item	Semantic Relations	Percentage of Correct Answers	Remarks
24	Group (Joint) Group (Separate)	62.56%	6 pupils made computational error
26	Group (Joint) Group (Separate)	58.71% 63.58% *(including those with computational errors)*	19 pupils made computational error from inability to rename (+0.0487)
30	Group (Joint) Group (Separate)	37.18%	3 pupils made computational error
27	Group (Joint) Vary (Equal groups — Multiplicative)	29.23%	4 pupils made computational error of $9 \times 9 = 18$
28	Vary (Equal Groups — multiplicative) Comparison	19.49%	
29	Vary (Equal groups — quotation) Group (Separate)	6.67%	
31	Comparison Group (Joint)	21.28%	

First, consistent with Yeap and Kaur (2001), problems involving more than one semantic relation (multi-related) yielded poorer performance than those of common semantic relation (single-relation) except for Item 30. The absence of an action verb for the second month (comparing the phrase "sold 895 in the first month" to "1042 bottles in the second month") might have contributed to pupils' inability to decide what might happen to the 1042 bottles in the second month thus explaining 13 pupils who added 1042 to the whole.

Second, the computational proficiency of the pupils affected their ability to arrive at the correct answer. Errors were found in simple addition and subtraction, addition and subtraction with renaming and retrieval of multiplication facts. Apparently, putting the arithmetic operations in context seems to increase pupils' chances of committing computational errors.

Third, pupils' responses showed a lack of understanding when either more than one-step or more than one operation is required to solve the problem. Pupils were likely to pluck out all the given numbers in accordance to their appearance in the problem statements and then applied the preferred operation which was addition in most cases. Other relevant information was ignored. There was an apparent lack of comprehension of the information they read. Geary (1996) explained that reading skills is one of the factors that influence the developmental change of one's problem solving ability. In fact reading skills affects general comprehension skills which determine one's ability to correctly represent problems. Newman (1983) added that even with an ability to comprehend the problem, translating accurately the information into forms of mathematical representation, which she termed transformation, affects pupils' ability to choose correctly the accompanying procedural and operational steps.

5.2 *Fractions*

The benchmark test on fractions (see Appendix D2) assesses pupils' ability to interpret and represent fractions; identify equivalent fractions; compare and order fractions; add and subtract fractions, and solve word problems. The pupils were most successful in addition and subtraction of like fractions suggesting that many pupils might have memorised the procedures without even understanding the basic fraction concepts.

5.2.1 Interpret and represent fractions

Table 17a
Items on Interpret and represent fractions — Distribution of common responses

Item	Response	n (%)
1. The circle is cut along the line. 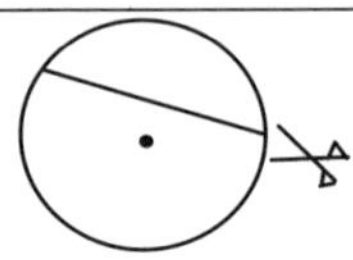	No*	304 (77.90)
	Yes	44 (11.28)
	No response	20 (5.13)
a) Has the circle been cut into 2 halves?		
b) Why? You may use diagram to explain.	Acceptable*	111 (28.46)
	Not acceptable	238 (60.87)
	No response	42 (10.74)
2. Share this bar of chocolate equally among 5 children. 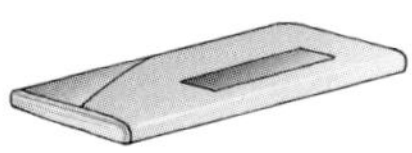 What fraction of the chocolate bar does each child get?	$\frac{1}{5}$ *	187 (47.95)
	$\frac{5}{5}$	59 (15.12)
	$\frac{2}{5}$	23 (5.90)
	$\frac{1}{2}$	22 (5.64)
	$\frac{4}{5}$	12 (3.08)
	$\frac{3}{5}$	11 (2.82)
	$\frac{5}{1}$	10 (2.56)
	$\frac{5}{10}$	9 (2.31)
	No response	57 (14.62)

Table 17b

Items on Interpret and represent fractions — Distribution of common responses

Item	Response	n (%)
3. Shade in two-thirds of this rectangle.	Shade 8 squares*	128 (32.82)
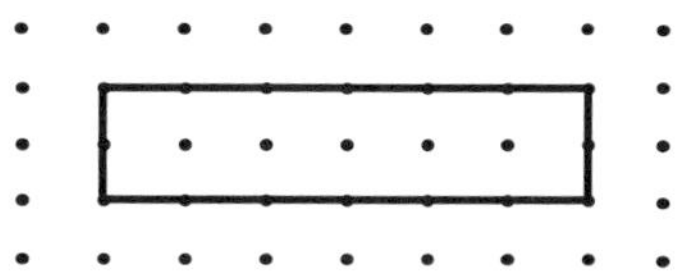	4 squares	74 (18.97)
	6 squares	64 (16.41)
	12 squares	21 (5.38)
	2 squares	20 (5.13)
	No response	50 (12.82)
4. What fraction of each of the shapes below is <u>shaded</u>? a)	$\frac{3}{4}$*	283 (72.56)
	$\frac{1}{4}$	40 (10.25)
	$\frac{1}{3}$	29 (7.44)
	$\frac{3}{1}$	16 (4.10)
b)	$\frac{2}{3}$*	277 (71.03)
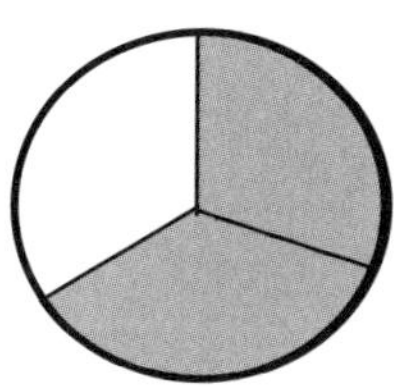	$\frac{1}{3}$	42 (10.77)
	$\frac{1}{2}$	25 (6.41)
	$\frac{2}{1}$	16 (4.10)

As shown in Tables 17a, about 78% of the sample of low attainers were aware that a chord does not cut the given circle in half (Item 1), but 30% were able to explain their reasoning. The drawing allows pupils to make the perceptual comparison of the two segments and to arrive at the correct answer. However many among them could not explain their thought. Those who could do so tend to refer to either the shape / size of the two segments or the centre of the circle. For example,

> *"When it is cut into 2 halves, we should cut into equal parts.*
> *You must cut (through) along the dote to cut into 2 halves.*
> *One of the side is small & the other side is big*
> *the left side of the circle has more space than the right side."*

Unacceptable responses include similar responses to those shown in Figure 5.

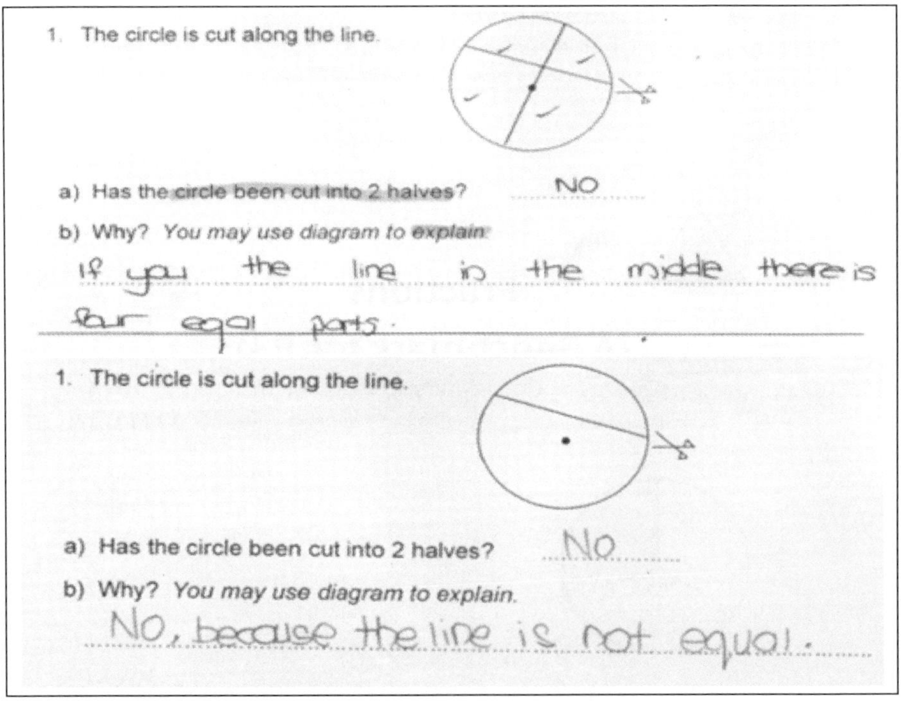
1. The circle is cut along the line.

a) Has the circle been cut into 2 halves? NO

b) Why? *You may use diagram to explain.*

if you the line in the midde there is four eqal parts.

1. The circle is cut along the line.

a) Has the circle been cut into 2 halves? No

b) Why? *You may use diagram to explain.*

No, because the line is not equal.

Figure 5. Samples of unacceptable responses for Item 1b

Pupils' lack of conceptual understanding of fraction was revealed when they were presented with a worthwhile task that incorporated $\frac{1}{5}$ (see Table 17a, Item 2). About two-third of the pupils had no idea at all what to do when asked to shade two-thirds of a given rectangle (see

Table 17b, Item 3). The square grid seemed to confuse them instead of helping them to divide the rectangle into three equal sections. However, the task of using a fraction label to represent the shaded region of a given diagram that was divided into equal sections is relatively easy (see Table 17b, Item 4) to the pupils. In Item 4, about 10% of the pupils labelled the unshaded region and 6% of the pupils incorrectly expressed the unshaded region as a fraction of the shaded region instead of the whole region.

5.2.2 Equivalent fractions

As reported in other studies (e.g. Hart, Brown, Kerslake, Küchermann, & Roddock, 1984), children find it very difficult to understand equivalence of rational numbers. In this study, more than 90% of the pupils performed poorly on items that tested their conceptual knowledge of equivalent fractions (see Table 18a, Items 7 and 8). They performed better when asked to identify the missing numerator / denominator in a given set of equivalent fractions.

Table 18a

Items on Equivalent fractions — Distribution of common responses

Item	Response	n (%)
7. Put a tick under the shape(s) that is shaded half.	(a) and (b)*	39 (10.00)
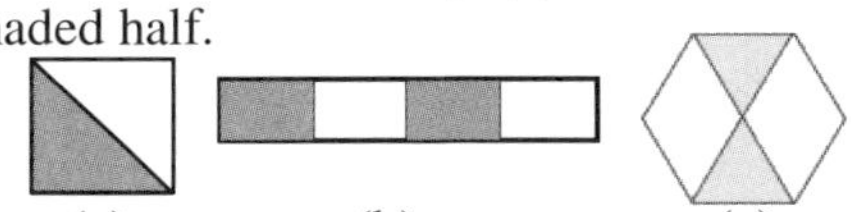	(a)	295 (74.36)
(a) (b) (c)	(a) and (c)	25 (6.41)
	(a), (b) and (c)	12 (3.08)
8. Meng and Ahmad each has a bar of chocolate of the same size. Meng breaks his into 6 equal pieces and eats 4 of them. Ahmad breaks his into 3 equal pieces and eats 2 of them.	Same amount*	17 (4.36)
a) Who eats more chocolate? _________	Meng	271 (69.49)
b) Why do you think so? You may use diagram to explain.	Ahmad	57 (9.66)
	No response	19 (4.88)

Table 18b
Items on Equivalent fractions — Distribution of common responses

Item	Response	n (%)
9. Fill in the missing numbers:	6*	199 (51.03)
a) $\frac{1}{3} = \frac{2}{?}$	3	132 (33.85)
	4	17 (4.36)
	1	16 (4.10)
	2	8 (2.05)
b) $\frac{6}{8} = \frac{3}{?}$	4*	171 (43.85)
	8	102 (26.15)
	16	34 (8.72)
	2	16 (4.10)
	6	12 (3.08)
	5	10 (2.56)
c) $\frac{6}{12} = \frac{?}{4}$	2*	159 (40.77)
	3	72 (18.46)
	6	38 (9.74)
	18	20 (5.13)
	12	17 (4.36)
	1	14 (3.59)
	7	14 (3.59)
	4	10 (2.56)
	5	9 (2.41)
	8	8 (2.05)
	No response	12 (3.08)
d) $\frac{2}{5} = \frac{?}{10}$	4*	171 (43.85)
	2	53 (13.59)
	1	47 (12.05)
	5	38 (9.74)
	3	22 (5.64)
	10	19 (4.87)
	7	11 (2.82)

As shown in Tables 18a and 18b, pupils found the concept of equivalent fractions puzzling. More than 74% of the pupils believed that the pictorial representation of half must only have two equal parts. It is not possible for a fraction to be represented by shapes divided into different number of equal parts. There were about 10% of the pupils who ignored the fact that the parts must be equal, and selected incorrectly the shaded hexagon as a picture that represented half.

In Item 8, pupils were given a task in which identical bars of chocolate were divided into different number of parts (6 and 3) and asked to compare 4 out of 6 with 2 out of 3, less than 4% of the pupils were able to give the correct answer. No drawing was included in the test item so that pupils' correct answer would not base on perception. Some pupils who gave the correct responses were able to use the idea of equivalent fractions (see Figure 6a) to explain their answer while others drew diagrams to illustrate their reasoning (see Figures 6b and 6c).

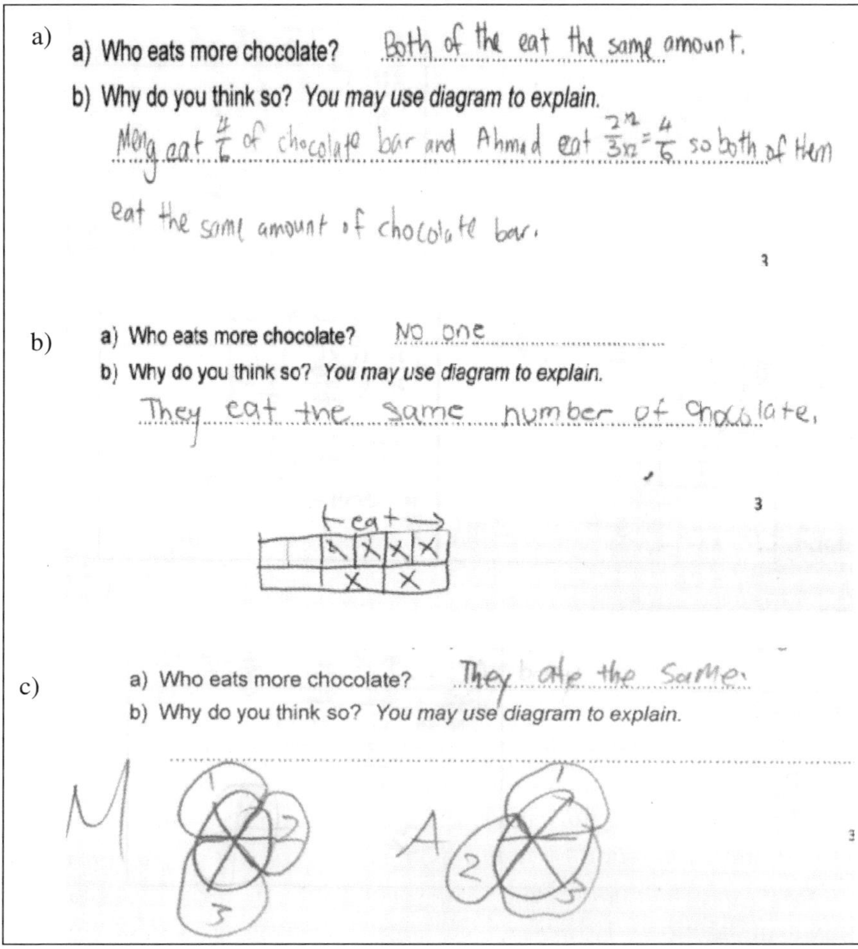

Figure 6. Sample of correct responses for Item 1b

Many pupils (about 46%) considered the number of parts and ignored the size of the parts in their comparison. In this case performance could be better if the task is put in division situation, a situation familiar to the children outside the classroom. An analogues division situation would be as follows:

4 boys share 6 chocolate bars and 2 girls share 3 chocolate bars. Who gets more chocolate, the boys or the girls?

In items testing pupils' procedural knowledge on equivalent fractions, as shown in Table 18b at least 40% of the pupils were able to perform the tasks using the "multiplying factor" technique or "dividing factor" technique to find the missing numerator / denominator in the given equivalent fractions. These techniques are shown in Figure 7.

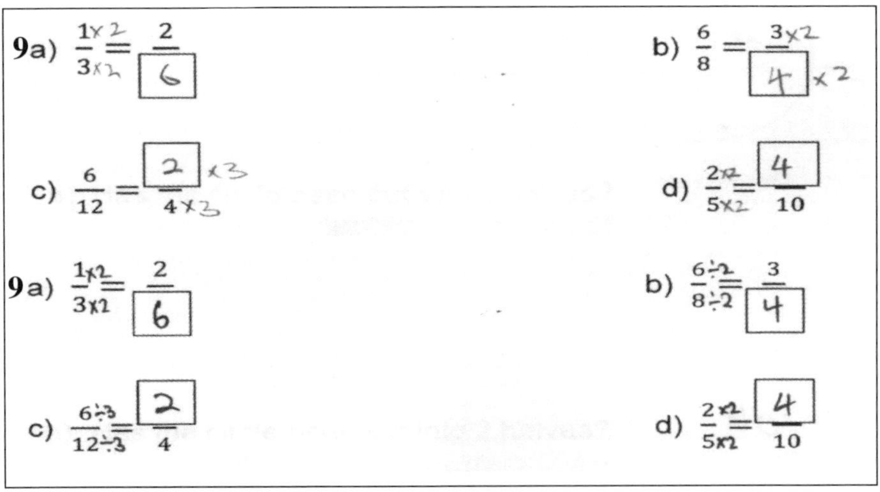

Figure 7. Sample of pupils' correct responses to Item 9

The most common incorrect response for the missing denominator is the given denominator, for example, 3 and 8 in Items 9a and 9b respectively. Pupils could have recalled one of the salient steps in addition and subtraction of fractions, that is, "*convert to fractions with the same denominator*", and overgeneralised the step to find the equivalent fractions. However, such misconception is not as prevalent when pupils were asked to find the missing numerator. The most common error occurs when pupils stopped prematurely at finding the quotient by dividing the two given denominators (see Figure 8). They did not multiply / divide the given numerator by the quotient. Other errors were probably due to the recall of incorrect steps like instead of multiply, they divided (see Figure 8).

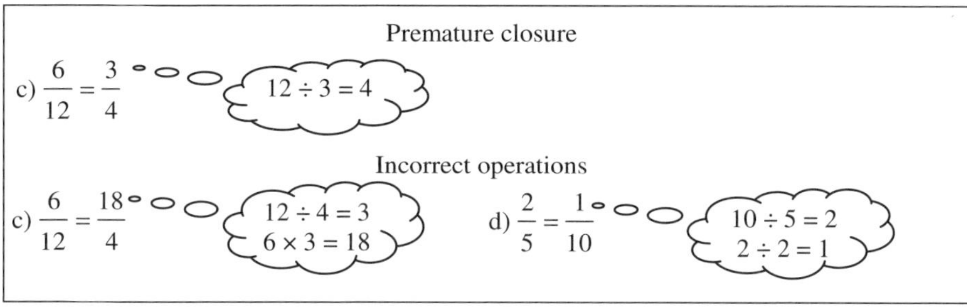

Figure 8. Sample of pupils' incorrect responses Item 9

5.2.3 Comparing and ordering fractions

It has been reported that both primary and secondary pupils have difficulties in comparing and ordering fractions (e.g., Hart, Brown, Kerslake, Küchermann, & Roddock, 1984; Nunes, Bryants, Pretzlik, & Hurry, 2006). As shown in Table 19, many Primary 4 low attainers had difficulty in comparing and ordering factions.

Table 19

Items on Comparing and ordering fractions — Distribution of common responses

Item	Response	n (%)
5. Circle the larger fraction in each of the pairs below. The first is done for you.		
a) $\frac{1}{4}$ (circled: $\frac{3}{4}$)		
b) $\frac{3}{7}$ $\frac{5}{7}$	$\frac{5}{7}$ *	338 (86.67)
	$\frac{3}{7}$	34 (8.72)
	No response	18 (4.62)
c) $\frac{3}{5}$ $\frac{3}{4}$	$\frac{3}{4}$ *	130 (33.33)
	$\frac{3}{5}$	239 (61.28)
	No response	21 (5.38)
d) $\frac{2}{5}$ $\frac{3}{10}$	$\frac{2}{5}$ *	148 (37.95)
	$\frac{3}{10}$	227 (71.02)
	No response	15 (3.85)
6. Put these fractions in order of size, from the largest to the smallest $\frac{1}{7}$ $\frac{1}{4}$ $\frac{1}{12}$	$\frac{1}{4}$ $\frac{1}{7}$ $\frac{1}{12}$ *	143 (36.67)
	$\frac{1}{12}$ $\frac{1}{7}$ $\frac{1}{4}$	204 (52.31)
	$\frac{1}{4}$ $\frac{1}{12}$ $\frac{1}{7}$	12 (3.08)

When asked to compare like fractions, about 87% of the pupils were able to answer correctly. Many of the pupils probably just considered the numerators (since the denominators are the same) and ordered them as if they were natural numbers. They applied the same strategy in Item 5c where they were asked to compare unlike fractions with the same numerators, and Item 5d where the unlike fractions have different numerators as well. As a result, the percentages of correct answers for Items 5c and 5d are about 33% and 38% respectively. The drop in the rate of correct responses is expected as pupils failed to consider the value of the fractions in a way different from comparing natural numbers. The incorrect responses in Items 5b and 5c indicate that pupils had not taken the numerator and the denominator as a representation for a single quantity. They failed to consider the relationship between the numerator and the denominator. Comparing and ordering fractions is different from comparing and ordering natural numbers. There are finite numbers between two different natural numbers but there are infinite fractions between two given fractions.

5.2.4 Adding and subtracting fractions

As shown in Tables 20a, 20b, 20c and 20d more than 85% of the pupils were able to add or subtract like fractions. They seemed to be able to recall the procedure of using the given denominator as the denominator in the answer and then adding / subtracting the numerators. However, when they were faced with different denominators, many had absolutely no idea as what to do. They were not aware of the connections between addition / subtraction of like fractions with that of the related fractions and the role played by equivalent fractions in the addition / subtraction algorithms.

The main source of the computational error is pupils' lack of understanding of fractions and its written representation. Pupils treated the written representation as two unrelated numerals separated by a line. They added / subtracted the numerators and the denominators separately. For example,

$$\frac{2}{7}+\frac{3}{7}=\frac{2+3}{7+7}=\frac{5}{14} \qquad \frac{7}{10}+\frac{1}{5}=\frac{8}{15} \qquad \frac{7}{8}-\frac{2}{8}=\frac{7-2}{8-8}=\frac{5}{0}$$

The other common error in the addition and subtraction of related fractions occurred when pupils added / subtracted the numerators but chose to use the larger denominator in the addends as the denominator for the sum / difference. For example,

$$\frac{7}{10}+\frac{1}{5}=\frac{7+1}{10}=\frac{8}{10} \qquad \frac{5}{6}-\frac{1}{6}=\frac{5-1}{6}=\frac{4}{6} \qquad \frac{2}{3}-\frac{1}{6}=\frac{2-1}{6}=\frac{1}{6}$$

Misinterpreting the operational symbols occurs only in the subtraction when pupils added the fractions instead of subtracted.

$$\frac{5}{6}-\frac{1}{2}=\frac{5}{6}-\frac{3}{6}=\frac{8}{6} \qquad \frac{2}{3}-\frac{1}{6}=\frac{4}{6}-\frac{1}{6}=\frac{5}{6}$$

Table 20a

Items on Addition and subtraction of fractions — Distribution of common responses

Item	Response	n (%)
10a) $\frac{2}{7}+\frac{3}{7}$	$\frac{5}{7}$ *	333 (85.39)
	$\frac{5}{14}$	23 (5.90)
10b) $\frac{7}{10}+\frac{1}{5}$	$\frac{9}{10}$ *	187 (47.95)
	$\frac{8}{15}$	76 (19.49)
	$\frac{8}{10}$	46 (12.05)
	$\frac{8}{5}$	12 (3.07)
	No response	9 (2.31)

Table 20b

Items on Addition and subtraction of fractions — Distribution of common responses

Item	Response	n (%)
10c) $\frac{1}{3}+\frac{1}{9}$	$\frac{4}{9}$*	204 (52.31)
	$\frac{2}{12}$	61 (15.64)
	$\frac{2}{9}$	41 (10.51)
	$\frac{2}{3}$	13 (3.33)
	$\frac{1}{12}$	11 (2.82)
10d) $\frac{3}{8}+\frac{2}{4}$	$\frac{7}{8}$*	188 (48.21)
	$\frac{5}{12}$	72 (18.46)
	$\frac{5}{8}$	54 (13.85)
	$\frac{4}{8}$	18 (4.60)
	$\frac{5}{4}$	10 (2.56)
11a) $\frac{7}{8}-\frac{2}{8}$	$\frac{5}{8}$*	339 (86.92)
	$\frac{5}{0}$	17 (4.36)

Table 20c
Items on Addition and subtraction of fractions — Distribution of common responses

Item	Response	n (%)
11b) $\frac{5}{6} - \frac{1}{2}$	$\frac{2}{6}$ or $\frac{1}{3}$ or $\frac{4}{12}$ *	180 (46.15)
	$\frac{4}{4}$	74 (18.98)
	$\frac{4}{6}$	37 (9.49)
	$\frac{4}{2}$	12 (3.08)
	$\frac{3}{6}$	9 (2.31)
11c) $1 - \frac{3}{5}$	$\frac{2}{5}$ *	234 (60.00)
	$\frac{2}{4}$	21 (5.39)
	$\frac{3}{5}$	15 (4.10)
	$1\frac{3}{5}$	11 (2.82)
	$\frac{1}{5}$	9 (2.31)
	No response	9 (2.31)

Table 20d

Items on Addition and subtraction of fractions — Distribution of common responses

Item	Response	n (%)
11d) $\frac{2}{3} - \frac{1}{6}$	$\frac{3}{6}$ or $\frac{1}{2}$ or $\frac{9}{18}$ *	181 (46.41)
	$\frac{1}{3}$	77 (19.74)
	$\frac{1}{6}$	47 (12.05)
	$\frac{5}{6}$	24 (6.15)
	$\frac{2}{6}$	8 (2.05)

5.2.5 Word problems

The percentages of correct answers for the word problems do not differ greatly from the items on the operations of fractions without context. However, pupils' inability to carry out the addition / subtraction of related fractions is not the only cause of error as shown in Tables 21a and 21b.

Table 21a

Items on Word problems — Distribution of common responses

Item	Response	n (%)
12. Maylin used $\frac{3}{5}$ kg of flour for bread and $\frac{3}{10}$ kg of flour for cakes. How much flour did she use altogether?	$\frac{9}{10}$*	192 (49.23)
	$\frac{6}{15}$	50 (12.82)
	$\frac{6}{10}$	38 (9.74)
	$\frac{3}{15}$	15 (3.85)
	$\frac{3}{10}$	13 (3.33)
13. Jane has $\frac{1}{6}$ m of ribbon. Nurul has $\frac{7}{12}$ m of ribbon.		
a) Whose ribbon is longer?	Nurul*	259 (66.41)
	Jane	23 (5.90)
b) How much longer?	$\frac{5}{12}$*	117 (30.00)
	$\frac{7}{12}$	36 (8.72)
	$\frac{9}{12}$	33 (8.46)
	$\frac{8}{12}$	19 (4.87)
	$\frac{8}{18}$	19 (4.87)
	$\frac{1}{6}$ or $\frac{2}{12}$	19 (4.87)
	$\frac{6}{6}$	13 (3.33)

Table 21b

Items on Word problems — Distribution of common responses

Item	Response	n (%)
14. There was some juice in the glass. Devi drank $\frac{1}{4}$ ℓ of the juice from the glass. There is $\frac{1}{8}$ ℓ of juice left in the glass. How much juice was there in the glass at first?	$\frac{5}{8}$	164 (42.05)
	$\frac{4}{12}$ or $\frac{1}{3}$	43 (10.77)
	$\frac{4}{8}$	30 (7.69)
	$\frac{2}{4}$	29 (7.44)
	$\frac{1}{8}$	27 (6.92)
	$\frac{2}{8}$	10 (2.56)
	No response	8 (2.05)

There are pupils who misinterpreted the problem statements and consequently chose the incorrect operation. This happens more often with the "compare" problem (see Table 21a, Item 13) than the "join" problem (see Table 21a, Item 12) and the "separate" problems (see Table 21b, Item 14). The two common computational errors found in Items 10 and 11 can also be found here e.g., add / subtract the numerators and add / subtract the denominators (e.g. $\frac{6}{15}$ in Item 12, $\frac{6}{6}$ in Item 13 and $\frac{4}{12}$ in Item 14) and add / subtract the numerators and keep the larger given denominator (e.g. $\frac{6}{10}$ in Item 12, $\frac{4}{8}$ in Item 14)

Item 12 requires pupils to add two related fractions with the same numerator 3. About 7% of the pupils decided to take 3 as the numerator of the sum of the two fractions. They gave incorrect answers such as these:

$$\frac{3}{5}+\frac{3}{10}=\frac{3}{15} \qquad \frac{3}{5}+\frac{3}{10}=\frac{3}{10}$$

In Item 13a, about 66% of the pupils could identify the longer length but only about 30% of them were able to find the difference in lengths. The relatively high percentage of correct responses for Item 13a may be due to that fact that some pupils could have compared the numerators and denominators separately and arrived at the fact that $\frac{7}{12}$ is greater.

5.3 *Measurement*

The benchmark test on Measurement, comprising two parts, assesses what the pupils were expected to know by the end of Primary 3. The test is shown in Appendix D3. In this section we report the results of both parts of the test according to the attributes: length, capacity, mass, area and perimeter and money.

5.3.1 Length

Length is the first attribute to be introduced in Primary 1. Even though the tasks in Item 1, as shown in Table 22a, are similar to a task found in the Coursebook for Primary 1, less than 63% of the pupils were able to give the correct responses. More pupils would be successful if they knew the correct labelling for a straight line. For example, some pupils labelled the line GH as "G and H" or "G, H".

As shown in Table 22b, the correct response is about 92% for Item 2a and only about 7% for Item 2b showing that pupils had great difficulty reading scale that does not have "0" as the starting point. They did not understand the scale, and did not realise that any point can serve as the origin (start) of a scale. This poor performance confirmed the finding reported in the large scale study in America by Sowder, Wearne, Martin, and Strutchens (2004). Less than 25% of the 4th graders and only about 60% of the 8th graders in their study were able to measure correctly in a task similar to that in Item 1b. Bryant (2007) suggested that the source of such errors may be the failure of the pupils to establish one-to-one correspondence between the number of units in the part of the ruler and the "imagined" units on the line itself.

For Item 3 as shown in Table 22b, less than 46% of the responses are correct. Many of the low attainers showed lack of understanding of the inverse relationship in units. They did not realise that as a unit increases in size, fewer units are required to measure the same length, an idea taught in Primary 1. However, pupils performed relatively better when asked to convert a measure in compound units (m and cm) to the smaller unit, and vice versa. The correct responses for Items 6d and 6b are 77% and 55% respectively. Conversion is taught in Primary 3 and involves mainly procedural knowledge.

Table 22a

Items on Length — Distribution of common responses

Item	Response	n (%)
1. Compare the lengths of the lines AB, CD, EF and GH drawn on the grid paper below.	a) GH*	a) 230 (58.97)
	CD	65 (16.67)
	EF	56 (14.36)
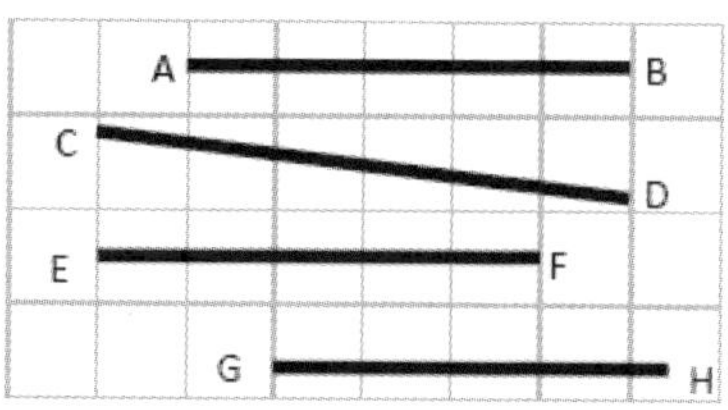	b) EF*	b) 249 (63.85)
	CD	65 (16.67)
	GH	42 (10.77)
	c) CD*	c) 238 (61.03)
	GH	91 (23.33)
	EF	27 (6.92)
a) Line AB is longer than line _______.		
b) Line AB is as long as line _______ .		
c) Line AB is shorter than line ______ .		

Table 22b

Items on Length — Distribution of common responses

Item	Response	n (%)
2. What is the length of each line?		
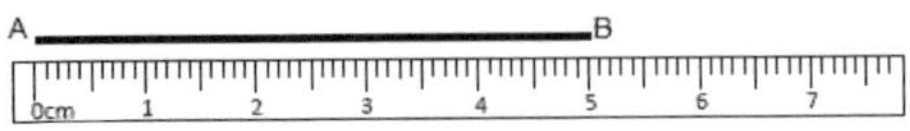	a) 5*	a) 358 (91.80)
a) Length of line AB is _____cm.		
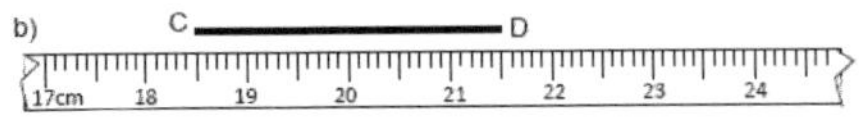	b) 3 or 3.05*	b) 29 (7.44)
	25	47 (12.05)
b) Length of line CD is _____cm.	21 or 21.05	44 (11.28)
	20	32 (8.21)
	26	30 (7.69)
	6	23 (5.90)
	4	23 (5.90)
	21.5 or 21.6	21(5.39)
	22	19 (4.87)
	7	18 (4.62)
	5	14 (3.59)
	30	10 (2.56)
3. Salma measures the width of the door using these different objects:	a) 38*	178 (45.64)
(i) paper clips (ii) matches (iii) cm cubes	19	65 (16.67)
	95	51 (13.08)
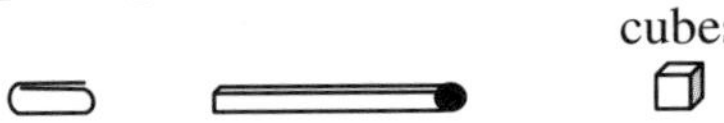	b) 19*	141 (36.15)
	95	88 (22.56)
She gets the answers 95, 38 and 19.	38	67 (17.18)
Which number goes with each object?		
a) Number of paper-clips used: ______	c) 95*	157 (40.26)
b) Number of matches used: ______	19	91 (23.33)
c) Number of cm cubes used: ______	38	48 (12.31)
6b) 405 cm = _____m _____cm	b) 4m 5cm*	217 (55.64)
	40m 5cm	69 (17.69)
	400m 5cm	36 (9.23)
	0m 405cm	17 (4.36)
6d) 3 m 20 cm = _______ cm	d) 320 cm*	302 (77.44)
	3020cm	37 (9.49)
	20cm	15 (3.85)
	23cm	9 (2.31)

5.3.2 Capacity and mass

Capacity, not volume, is taught at the lower primary level. The correct responses to tasks that required pupils to convert compound units (ℓ and mℓ) to smaller unit (mℓ), and vice versa, are about 84% and 62% respectively (see Table 23). Apparently, pupils had more difficulty with the conversion of mass than capacity as less than 50% of the pupils were able to convert mass in smaller unit (g) to compound unit (kg and g), and only 25% of the pupils were able to compare the masses in Item 7.

Table 23
Items on Capacity and mass — Distribution of common responses

Item	Response	n (%)
6. Fill in the blanks.	2 kg 650g*	193(49.49)
a) 2650 g = _____ kg _____g	26kg 50g	97 (24.87)
	2000kg 650g	30 (7.69)
	265kg 0g	8 (2.05)
c) 7400 mℓ = ____ ℓ ___mℓ	7 ℓ 400 mℓ*	240 (61.54)
	74 ℓ 00 mℓ	43 (11.03)
	7000 ℓ 400 mℓ	28 (7.18)
	7 ℓ 4 mℓ	11 (2.82)
e) 6 ℓ 507 mℓ = ________ mℓ	6507 mℓ	327 (83.85)
	507 mℓ	14 (3.59)
	No response	8 (2.05)
7. Shade the <u>lightest</u> mass.	2kg 150g*	99 (25.39)
	22kg	158 (40.51)
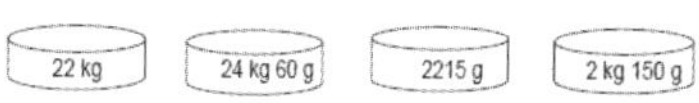	2215g	94 (24.10)
	24kg 60g	11 (2.82)
	No response	17 (4.36)

To answer Item 7 correctly, pupils need to convert 2215g to 2 kg 215g and then compare it with 2kg 150g. However, some pupils converted incorrectly all four measures to the smaller unit (g) and then compared, creating more opportunities for them to commit processing errors when they had to compare 4-digits and 5-digit numbers. About 40% of the pupils omitted the units in the four measures and compared the numerical values of 22, 2460, 2215 and 2150 instead (see Figure 9).

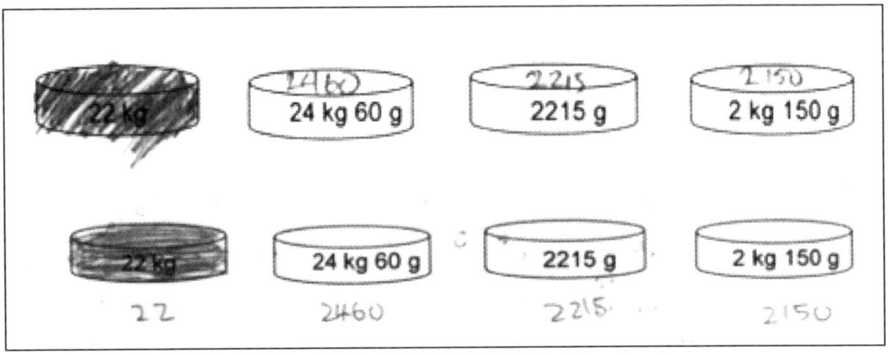

Figure 9. Sample of pupils' incorrect responses to Item 7

5.3.3 Area and perimeter

Pupils found it difficult to compute the perimeter of the shapes in Item 4 (see Table 24). Many of them seemed to focus on the salient features of the shapes instead. About 26% of the pupils believed that Shape X had the same perimeter as Shape A since both shapes shared a common feature; a reflex angle. Another 25% of the pupils chose Shape Y. They probably thought that Shape Y was similar to Shape A as both shapes could be "enclosed" in a 4 by 2 rectangle. About 7% of the pupils chose both Shape X and Shape Y because like Shape A, they are concave figures. There are also 5% of the pupils who counted the diagonal of the grid incorrectly as 1 unit, and hence concluded that both Shape Y and Z had the same perimeter as Shape A, that is, 12 units.

Table 24

Items on Area and perimeter — Distribution of common responses

Item	Response	n (%)
4. Circle the shape which has the same <u>perimeter</u> as Shape A.	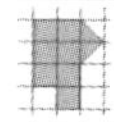 *	60 (15.39)
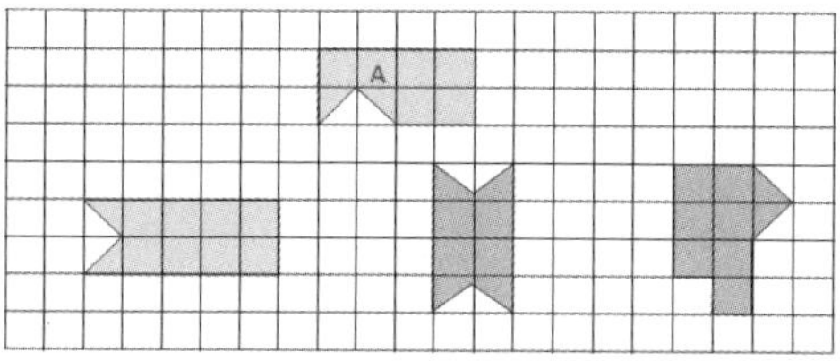	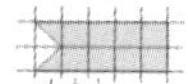	103 (26.41)
	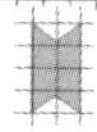	98 (25.13)
	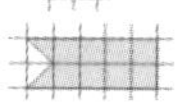 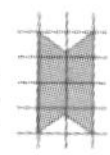	28 (7.18)
Shape X Shape Y Shape Z	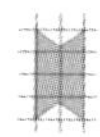 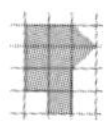	21 (5.39)
(Note: the figures will be referred to as Shapes X, Y and Z in the report.)	No response	72 (18.46)
5. Find the area of each shape in square centimeters.	a) 10*	163 (41.8)
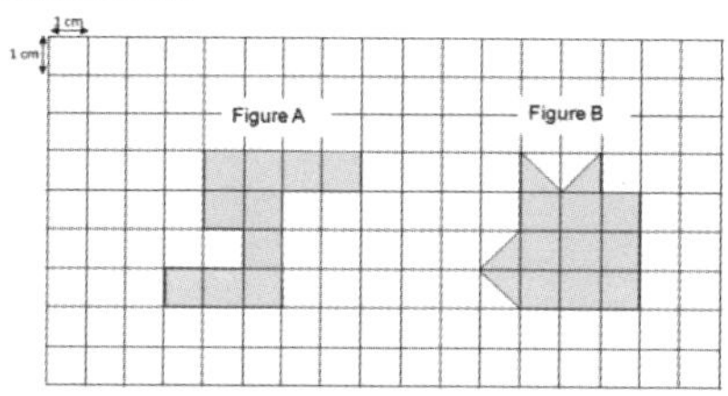	20	73 (18.72)
	19	19 (4.87)
	4	14 (3.59)
	40	12 (3.08)
	b) 11*	126 (32.31)
	13	58 (14.87)
	14	42 (10.77)
	10	20 (5.13)
a) Area of Figure A = ______ cm^2	12	19 (4.87)
b) Area of Figure B = ______ cm^2	9	16 (4.01)
	2	13 (3.33)
	16	10 (2.56)
	No response	9 (2.3)

Pupils performed relatively better when asked to find the area of a rectilinear figure (see Table 24, Item 5a) than that of a composite figure made up of a square and four triangles (see Table 24, Item 5b). The most common error is when the pupils confused the concept of area with the concept of perimeter. They counted the unit length around the given figure to determine the area i.e., 20cm^2 for Item 5a and 14cm^2 for Item 5b (see Figures 10a and 10b). For Item 5b, about 15% of the pupils counted the half square as 1cm^2 as shown in Figure 10c.

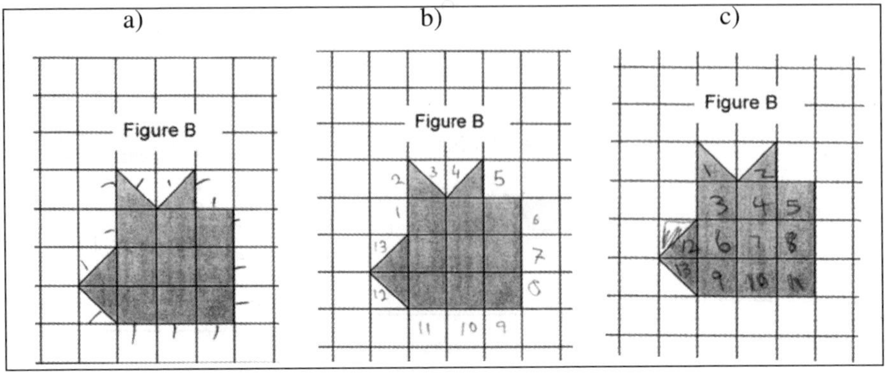

Figure 10. Sample of pupils' incorrect responses for Item 5

5.3.4 Money

As seen in Table 25 pupils performed relatively better in the task that required them to write the given amount of cents in decimal notation than vice versa.

Table 25

Items on Money — Distribution of common responses

Item	Response	n (%)
8. Fill in the blanks.	3720¢	277 (71.03)
a) \$37.20 = _____ ¢	20¢	39 (10)
	37.2¢	17 (4.36)
b) 3665¢ = \$ ______	\$36.65	300 (76.92)
	\$3665	24 (6.15)
	\$36	8 (2.05)
	No response	8 (2.05)

The difficulty here seems to be the interpretation of the role of the point in the decimal notation. 10% of the pupils thought that Item 8a required them to identify the cents. They identified correctly the 20 in \$37.20 as representing the cents and ignored the equality sign. 20 in \$37.20 represent the cents but it is not equivalent to \$37.20. These pupils lacked the understanding of equivalence in the representation. Such learning difficulty also led some pupils to write 3665¢ = \$ 36.

5.3.5 Word problems

The common learning difficulties in solving word problems are (a) misunderstand the problem statements and consequently translate them incorrectly into mathematical sentences, and (b) perform incorrectly the calculations. Of these two sources of error, misinterpretation of the problem statements was found to be more prevalent among the pupils, as shown in Table 26, and led them to

- read the scale instead of finding the mass of a packet of rice in Item 9,
- find the length of a pole that is 3 times as long as instead of 3m longer in Item 10,
- find the total amount of water in the pail and the tank in Item 11,
- add when the word "more" or subtract when the word "less", is present in the problem statements in Items 12, 13 and 14.

Table 26
Items on Word problems — Distribution and common responses

Item	Response	n (%)
9. The packets of rice have the same mass. Find the mass of a packet of rice.________kg 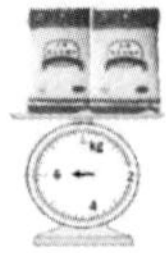	3kg*	195 (50.00)
	6kg	117 (30.00)
	4kg	24 (6.15)
	2kg	15 (3.85)
	12kg	12 (3.08)
10. A bamboo stick is 200 cm long. A lamp pole is 3 m longer. How long is the lamp pole? _____ cm	500cm*	47(12.05)
	600cm	103 (26.41)
	3200cm	93 (23.85)
	197cm	28 (7.18)
	300cm	26 (6.67)
11. A pail holds 8 litres of water. A tank holds 48 litres of water. How many pails of water are needed to fill the tank? ______litres	6*	149 (38.21)
	56	70 (17.95)
	40	43 (11.03)
	384	34 (8.72)
12. Lily has $57.80. Siti has $2 more than Lily. How much does Siti have? $______.	$59.80*	204(52.31)
	$57.82	45 (11.54)
	$55.80	26 (6.67)
	$115.60	17 (4.36)
	$57.78	13 (3.33)
13. Ahmad has $88.65. Meng has $28.60. a) Who has more money? ______. b) How much more? $ ______.	a) Ahmad*	a) 299 (76.67)
	117.25	23 (5.90)
	60.05	16 (4.10)
	Meng	6 (1.54)
	b) $60.05*	b) 192 (49.23)
	117.25	43 (11.03)
	88.65	17 (4.36)
	6005	11 (2.82)
14. Joshua has $16.80. He has 60 cents Less than Tim. a) How much does Tim have? $___. b) How much money do they have altogether? $______.	a) $17.40*	a) 95 (24.36)
	16.20	194 (49.74)
	1620	10 (2.56)
	b) $34.20*	b) 68 (17.44)
	$33	136 (34.87)
	$17.40	22 (5.64)
	$16.80	14 (3.59)
	$17	10 (2.56)

Most pupils might have mastered the operations on whole numbers but they had difficulties in the addition and subtraction of money using the decimal notation taught in Primary 3. (Note: the concepts of decimal and operations on decimals are introduced in Primary 4). Some pupils aligned the addends to the right instead of aligning the decimal points. For example in Item 12, about 12% of the pupils have understood the problem and chosen the correct operation but wrote

$$\begin{array}{r} 57.80 \\ +\quad 2 \\ \hline 57.82 \\ \hline \end{array}$$

There are also pupils who did not understand the decimal notation of money, thinking that \$16.20 can be written as \$16.2.

5.4 *Geometry*

Geometry can be considered as the mathematics of space in primary school mathematics. The main topics taught in lower primary include lines, shapes and patterns, and angles. In this section, pupil performance in the benchmark test on geometry, shown in Appendix D4, will be discussed under three headings: lines, shapes, and angles.

5.4.1 Lines

The results in Table 27 show that pupils seemed to find it easier to distinguish between straight lines and curved lines than between parallel lines and perpendicular lines. About 67% of the pupils were able to identify correctly a shape that is made up of 3 straight lines and a curved line (see Table 27, Item 1). The most common error occurs when pupils attempted to "decompose" a given shape into basic shapes and then to count the total number of straight edges and curve edges of the basic shapes (7%) as shown in Figure 11. If this misconception is not remediated and the pupils are tasked to find the perimeter of a composite shape, they would probably find the sum of the perimeter of the basic shapes that make up the composite shape.

Table 27
Items on Lines — Distribution of common responses

Item	Response	n (%)
1. Bala chose a shape that is made up of 3 straight lines and one curved line.	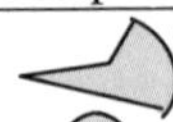 *	262 (67.18)
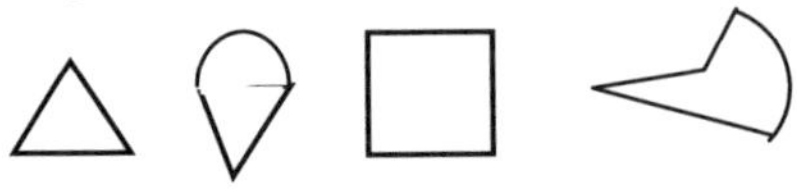		28 (7.18)
		8 (2.05)
Colour the shape that he chose.		7 (1.80)
3. Circle the letter(s) below that has a pair of parallel lines. T K H X	H*	250 (64.1)
	X	44 (11.28)
	T	28 (7.18)
	K	18 (4.62)
	T & H	11 (2.83)
	H & X	10 (2.56)
	No response	10 (2.56)
4. Which lines in the figure below are perpendicular? Draw along these perpendicular lines.	Lines 2 & 3*	144(36.67)
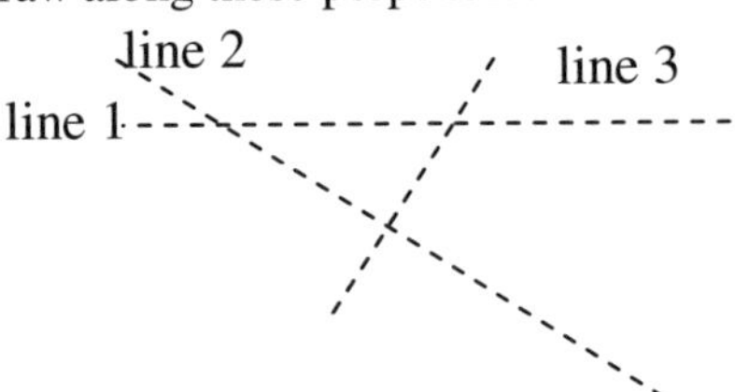	Lines 1, 2 & 3	70 (17.95)
	Lines1 & 2	57 (14.62)
	Line 1	26 (6.67)
	Line 3	21 (5.39)
	Line 2	16 (4.10)
	Lines 1 & 3	15 (3.85)
	No response	41 (10.51)

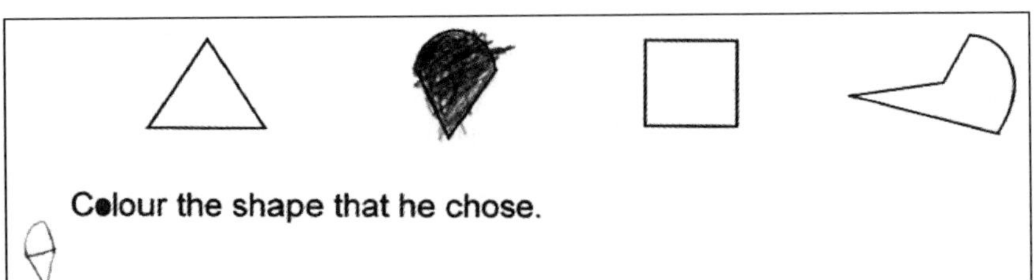

Figure 11. Misconception of finding the sum of the basic shape for composite shape

There are abundant examples of parallel lines and perpendicular lines in the children's spatial world outside the classroom, however only about 64% and 37% of the pupils were able to identify the parallel lines (see

Table 27, Item 2) and perpendicular lines (see Table 27, Item 3) respectively. These two geometric concepts are often taught in a way divorced from "reality", teachers tend to talk about parallel lines as "lines that never meet" and perpendicular lines as "lines that meet at right angle". The parallel lines and perpendicular lines are often represented as follows:

Consequently, many pupils did not think that the following can be parallel lines or perpendicular lines illustrating classic "orientation" learning difficulties.

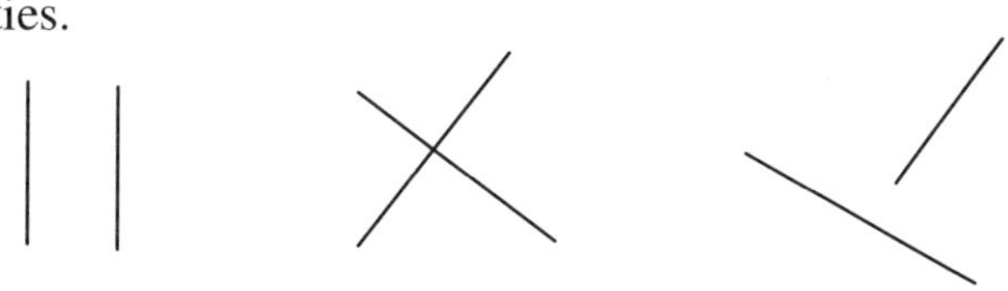

5.4.2 Shapes

It is apparent from Table 28 that some pupils had difficulties with the labelling of 2-D shapes in Item 2 and the solids in Item 9. In Item 9, about 92% of the pupils were able to identify the cubes but only about 66% of them were able to identify the cuboids.

For patterns involving the basic shapes, about 90% of the pupils were able to find the missing shape in an AB pattern as found in Item 7. The common error was committed by pupils who had chosen the shape correctly but ignored colour, the other attribute. Only about 66% of the pupils were able to answer Item 8 correctly. Here, a more complex pattern is given which is a composition of two patterns: AB (for shapes) and AAB (for shading). About 26% of the pupils selected the correct shape but not the correct colour.

Table 28
Items on Shapes — Distribution of common responses

Item	Response	n (%)
2. Circle the name of the shape that has four sides and four corners. A triangle A circle A half circle A rectangle	A rectangle*	308 (78.97)
	A triangle	27 (6.92)
	A half circle	27 (6.92)
	A circle	11 (2.82)
	No response	9(2.31)
7. Look at the pattern. 	*	353 (90.51)
		14 (3.59)
Circle the missing shape.		6 (1.54)
	No response	11 (2.82)
8. Look at the pattern. 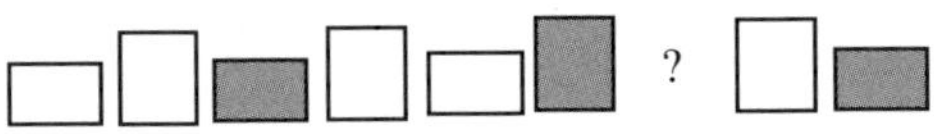	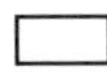*	261 (66.92)
		104 (26.67)
Circle the missing shape. 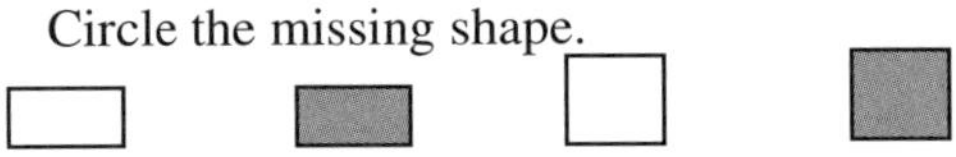	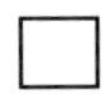	9 (2.31)
	No response	10 (2.56)
9. Peter built this structure. He used, 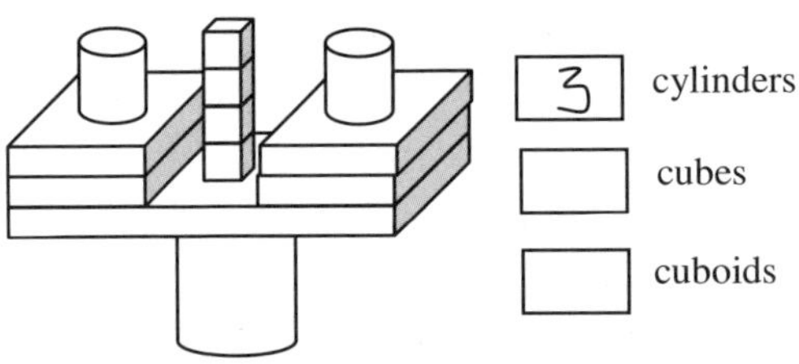	3, 4, 5*	259 (66.41)
	3, 4, 4	46 (11.80)
	3, 4, 2	17 (4.36)
	3, 4, 1	12 (3.08)
	3, 4, 6	12 (3.08)
	3, 4, 3	311 (2.82)
	3, 2, 4	10 (2.56)

5.4.3 Angles

As shown in Table 29, the concept of angle is a relatively difficult concept for the pupils. In addition, pupils had no idea of what it means by equal angles. More than 43% of the pupils circled only one of the angles when asked to circle two equal angles in Item 5. Only about 5% of the pupils committed the error of identify two angles with sides of equal length as equal angle.

The percentages of pupil who could identify the angle greater or smaller than a right angle in Item 6 were about 69% and 54% respectively. There were also many pupils (about 13%) who did not attempt this item. The most common incorrect response is the choice of the right angle. Item 6 was a difficult task to these pupils probably because the task required them to consider the relation between the orientation of the two lines with that of a right angle and the lines of the given angles are not horizontal.

Table 29

Items on Angles — Distribution of common responses

Item	Response	n (%)
5. Circle the two equal angles.	1 & 3*	160 (41.03)
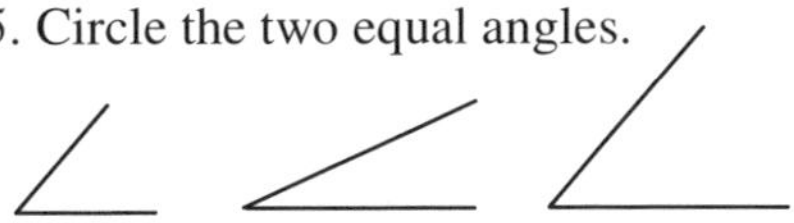	1 or 2 or 3	169 (43.33)
	1 & 2	28 (7.18)
	2 & 3	21 (5.39)
(denoted by Figures 1, 2, 3 respectively for easy reporting.)		
6. Write the letter A on the angle that is larger than a right angle. Write the letter B on the angle that is smaller than a right angle.	Marked as larger than a right angle:	
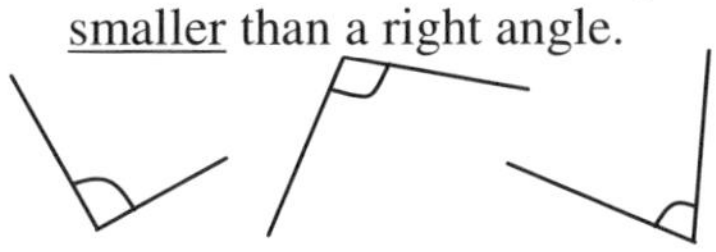	2*	269 (68.97)
	1	34 (8.72)
	1& 2	14 (3.59)
	3	12 (3.08)
	No response	52 (13.33)
(denoted by Figures 1, 2, 3respectively for easy reporting.)	Marked as smaller than a right angle:	
	3*	209 (53.59)
	1	84 (21.54)
	1 & 3	31 (7.95)
	2	9 (2.31)
	No response	52 (13.33)

5.5 *Data analysis*

The test items on data analysis, shown in Appendix D5, can be categorised under three headings: reading the data, reading between the data, and drawing the missing bar in a bar graph. The first two categories correspond to the first two levels of graph comprehension. Reading beyond the data, the third level of graph comprehension that involves eliciting and evaluating opinions about the information represented was not assessed in this test. The third category tests on pupils ability to represent missing information in the given graph.

5.5.1 Reading the data

As expected, pupils performed relatively well in items that assessed this lowest level of graph comprehension. The percentages of correct responses range from 69% to 93% as shown in the Table 30. The main learning difficulties at this level of graph comprehension include:

- difficulty with the scales: In Item 1a, nearly 21% of the pupils counted the icons in the correct bar but did not multiply the number by 2.
- misunderstanding the items: In Item 1c, when asked the task to identify the number of different pets P3A pupils have, about 8% of the pupils gave the answer of 30 or 15. That is, the total number of pupils who have pets and the total number of icons respectively.
- difficulty with zero data. In Item 3a, more than 6% of the pupils did not consider green (with value zero) as a valid category in the bar graph, and in Item 3c, about 9% of the pupils considered zero as an unacceptable value for green M&M's, hence gave 1 as the answer instead.
- focusing on the quantitative aspects of the graph and did not consider the spatial elements of the given graph. For example, there are five bars in the graph in Item 3 and so 5 is the most common incorrect response for Item 3a, pupils had ignored the wide space between the bars representing the number of blue candies and the brown candies. They had concentrated on the perceptually salient features of the

representation and not attending to all the elements of the graph given.

Table 30

Items on Reading the data — Distribution of common responses

Item	Response	n (%)
1. In P3A, every pupil has a pet. Look at the graph below. 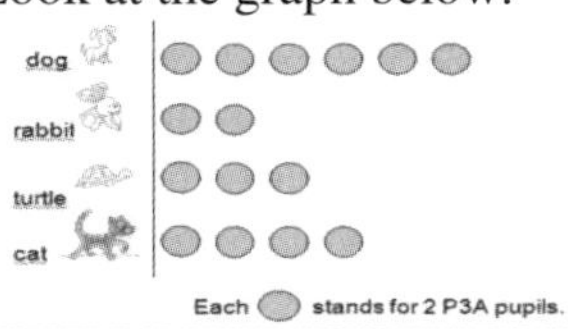		
a) How many pupils have dogs?	12*	297 (76.15)
	6	81 (20.77)
c) How many different pets do P3A pupils have?	4*	271 (69.49)
	30	22 (5.64)
	2	15 (3.85)
	15	11 (2.82)
3. The graph shows the favourite M&M's colours of pupils from P4A. 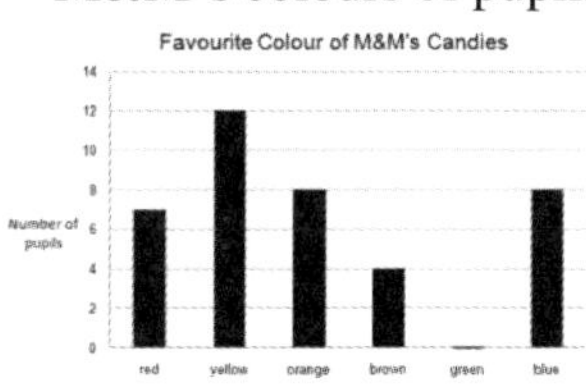		
a) How many different colours of M&M's candies are there?	6*	283 (72.56)
	5	26 (6.67)
	39	16 (4.10)
b) Which is the most popular colour?	Yellow*	364 (93.33)
c) How many pupils like green M&M's?	0*	304 (77.95)
	1	37 (9.49)

5.5.2 Reading between the data

Table 31 shows that the percentages of correct answers in this category of items decrease sharply from those in the previous category. They range from 48% to about 58%, indicating that about half of the pupils had difficulties reading between the data given in the graph.

Table 31

Items on Reading between the data — Distribution of common responses

Item	Response	n (%)
1. In P3A, every pupil has a pet. Look at the graph below. 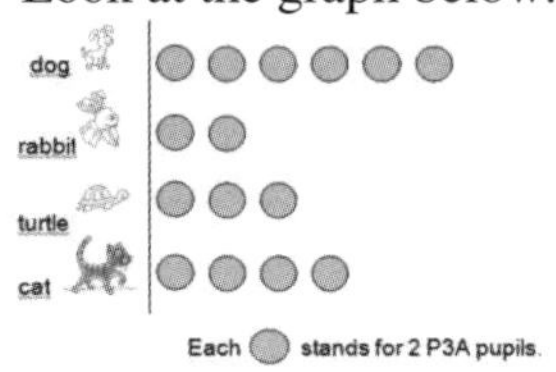		
b) How many more pupils own dogs than cats?	4*	209 (53.59)
	20	56 (14.36)
	10	40 (10.26)
	2	37 (9.49)
	8	12 (3.08)
d) How many pupils are there in P3A	30 *	225 (57.69)
	15	62 (15.90)
	4	18 (4.62)
3. The graph shows the favourite M&M's colours of pupils from P4A.	1	190(48.72)
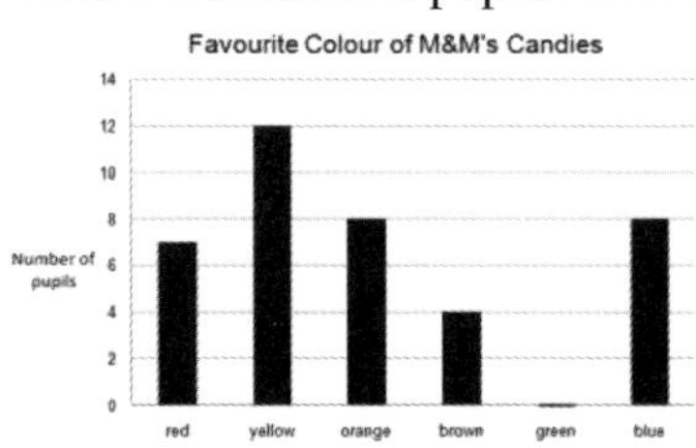	15	53 (13.59)
	8	36 (9.23)
	2	22 (5.64)
	5	10 (2.56)
	7	10 (2.56)
	Colours	16 (4.10)
d) How many more pupils like orange M&M's than red M&M's?		

The three main learning difficulties are as follow:

- pupils searched for the comparison "key" word and used it to determine the arithmetic operation. Both Items 1b and 3d have the word "more" in the problem statements. More than 24% and 13% of the pupils chose to add the correctly identified values instead of subtract in Items 1b and 3d respectively. Apparently, they were not aware that the items required them to find the "difference" and not the "larger" set in these compare situations.

- pupils had difficulties with the scales. In Item 1d, about 16% of the pupils did not multiply the total number of icon (15) by 2.
- pupils did not understand the task. There were pupils who answered item that required numerical answer with the category label and vice versa.

5.5.3 Completing the bar graph

To be able to insert the missing bar in a given graph, pupils need to understand the problem statements and interpret correctly the incomplete bar graph given.

Pupils found it relatively more difficult to complete the bar graph than reading the data and reading between the data in the graph. As shown in the Table 32, about 50% of the pupils were able to draw the missing bar correctly for the bar graph in Item 2, the most common incorrect response was shading six cells instead of three cells. About 23% of the pupils ignored the scale; they did not attempt to figure out the scale for the given bar graph. About another 7% of the pupils simply shaded all the seven cells available in the column.

Item 4 is relatively more difficult. Less than 37% of the pupils were able to figure out that Bala's scores must be less than 63 (2nd place) and greater than 61 (4th place). 19% of the pupils probably examined the bars starting from the horizontal axis, taking 63 as the highest score (1st place), 61 as the next highest score (2nd place) and leaving Bala's scores as the third highest (3rd place). Counting backwards by 2 from 63, these pupils concluded that Bala's scores was 59 and proceeded to draw a horizontal bar at the correct place to represent 59. Another 10% of the pupils thought in a similar way but they did not count backwards by 2, they just drew a bar to represent a number less than 61, which is 60, in this case.

Table 32
Items on Completing the bar graph — Distribution of common responses

Item	Response	n (%)
2 The graph shows the number of birthday cards Jane received from different countries. She received 6 cards from Australia. Draw the missing bar in the graph to represent cards that she received from Australia.	*The number of cells shaded:*	
	3*	196 (50.26)
	6	91 (23.33)
	7	28 (7.18)
	No response	23 (5.90)
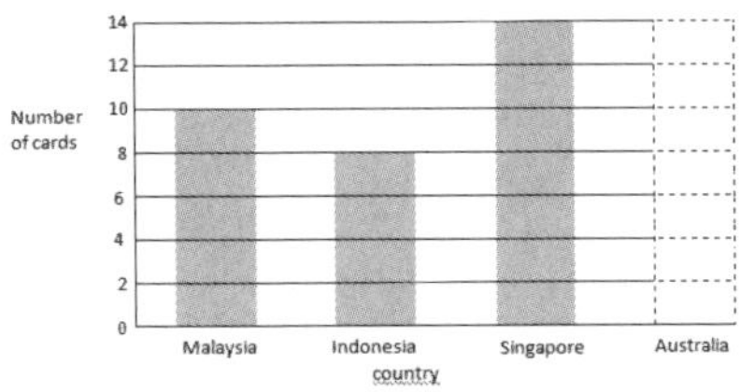		
4. This graph shows the scores obtained by 4 boys in a competition. Helmi is in the first place, Bala is in the third place. Draw a bar to represent the scores of Bala.	Shading the bar to show:	
	62*	142 (36.61)
	59	75 (19.23)
	60	40 (10.26)
	64	40 (10.26)
	63	26 (6.67)
	61	17 (4.36)
	58	13 (3.33)
	No response	28 (7.18)
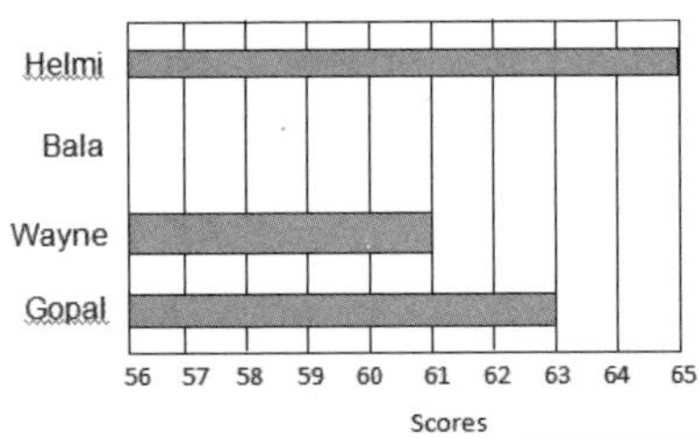		

6 Discussion and Concluding Remarks

The results revealed valuable information for teachers in the area of instructions, marking scheme and memory strategies. Low attainers often have similar learning difficulties than their peers, but they process information at a slower rate and tend to solve mathematical problems using strategies that differ in quality from the strategies used by their peers. Moreover, they tend to respond incorrectly in an assessment task not because they do not have the relevant knowledge, but, rather, because they are not able to access the information during the testing situation. They do not know what they are supposed to do because of their poor reading ability. These low attainers may perform better if the test item is read to them instead of they have to read the item themselves. Their lack of language skills to read and comprehend the test items and at times, explain their reasoning in words or using drawings, affects their test performance.

Appropriate language ability is necessary in solving word problems. Primary mathematics curriculum has a large component of word problems listed under various strands, from whole numbers to data analysis. Visual imagery can be used to solve most of these mathematical problems. Hegarty and Kozhevnikov (1999) distinguished between two types of visual-spatial representations used to solve mathematical problem: the pictorial representation and the schematics representation. The former encodes the visual appearance of the objects described while the latter encodes the spatial relations described in a problem and represents the essential information relevant to solving the problem. Schematic representations promote problem-solving success. Research has shown that pupils with learning difficulties tend to use fewer representation strategies while problem solving and the representations they used tend to be pictorial, compared to their peers (Montague & Applegate, 2000). In Singapore, a common visual representation used in solving word problems is the model. Most low attainers would attempt to draw models, often as pictorial representations (i.e., a rectangular bar), while solving problems. They need to be taught how to understand and interpret the problem statements and represent the relation elements found in the problem statements in the model they drawn. Classroom instruction should address these skill deficits of the low attainers in order to improve their performance in solving word problems.

With few exceptional cases, most of the low attainers in this exploratory study had the capability to learn. However, perpetual failures in school mathematics assessment might have created another obstacle in their mathematics education. Hence, teachers need to refine their instructional approach, set achievable learning goals and appropriate assessment tasks for these pupils. They should also teach these pupils explicitly the memory strategies and problem-solving strategies that would improve pupils' performance. Success would breed confidence and improve disposition towards mathematics. A confident pupil with a "I can" attitude would take mathematical problem solving in his / her stride.

References

Bishop, A.J. (1986). What are some obstacles to learning geometry. In R. Morris (Ed.), *Studies in Mathematics Education: Geometry in Schools, 5*, 141-159.

Bryant, D.P., Bryant, B.R., & Hammill, D.D. (2000). Characteristic behaviors of students with LD who have teacher-identified math weaknesses. *Journal of Learning Disabilities, 33*(2), 168-199.

Bryant, D.P. (2007). Key understanding in mathematics learning. Paper 5: *Understanding space and its representation in mathematics*. Nuffield Foundation.

Carpenter, T.P., Moser, J.M., & Bebout, H.C. (1988). Representation of addition an subtraction word problems. Journal for Research in Mathematics Education, 19(4), 345-357.

Diezmann, C. M., & Lowrie, T. (2006). Primary students' knowledge of and errors on number lines. In P. Grootenboer, R. Zevenbergen & M. Chinnappan (Eds.), *Proceedings of the 29th Annual Conference of the Mathematics Education Research Group of Australiasia* (Vol. 1, pp. 171-178) Sydney, Australia: MERGA.

Fischbein, E., Deri, M., Nello, M.S., & Marino, M.S. (1985). The role of implicit models in solving verbal problems in multiplication and division. *Journal for Research in Mathematics Education, 16,* 3-17.

Friel, S.N. Curcio, F.R., & Bright, G. (2001). Making sense of graphs: Critical factors influencing comprehension and instructional implications. *Journal for Research in Mathematics Education, 32*(2) 124-158.

Fuchs, L.S., Fuchs, D., Hamlett, W.L., Stuebing, K., & Fletcher, J.M. (2008). Problem solving and computational skill: Are they shared or distinct aspects on mathematical cognition? *Journal of Educational Psychology, 100* (1), 30-47.

Geary, D.C. (1996). *Children's mathematical development: Research and practical applications*. London, American Psychological Association.

Geary, D.C. (2004). Mathematics and learning disabilities. *Journal of Learning Disabilities, 37*(1), 4-15.

Gracia, A.L., Jimenez, J.E., & Hess, S. (2006). Solving arithmetic word problems: An analysis of classification as a function of difficulty in children with and without arithmetic LD. *Journal of Learning Disabilities*, *39*(3), 270-281.

Hart, K., Brown, M., Kerslake, D., Kuchermann, D., & Ruddock, G. (1984). *Chelsea diagnostic mathematics tests — Fractions 1*. Windsor (UK): NFER-Nelson.

Haylock, D. (1991). *Teaching mathematics to low attainers, 8-12*. London: Paul Chapman Publishing Ltd.

Hegarty, M., & Kozhevnikov, M. (1999). Types of visual-spatial representations and mathematical problem solving. *Journal of Educational Psychology, 91*(4), 684-689.

Jordan, N.C., Levine, S.C., & Huttenlocher, J. (1995). Calculation abilities of young children with different patterns of cognitive functioning. *Journal of Learning Disabilities,28*(4), 53-64.

Lowrie, T., & Diezmann, C.M. (2007). Solving graphics problems: Student performance in junior grades. *The Journal of Educational Research, 100*(6), 369-377

Ministry of Education. (2006). *Mathematics syllabus — Primary*. Singapore: Author.

Montague, M., & Applegate, B. (2000). Middle school students' perceptions, persistence, and performance in mathematical problem solving. *Learning Disability Quarterly, 23*, 215-227.

Newman, A. (1983). *The Newman language of mathematics kit.* Melbourne; Harcourt Brace Jovanovich Group.

Nunes, T., Bryant P., Pretzlik, U., & Hurry, J. (2006). *Fractions: Difficult but crucial inmathematics learning*. London Institute of Education, London: ESRC — Teaching and Learning Research Programme.

Petitto, A.L. (1990). Development of numberline and measurement concepts. *Cognition and Instruction, 7*, 55-78.

Sowder, J., Wearne, D., Martin, G., & Strutchens, M. (2004). What grade 8 students know about mathematics: Changes over a decade. In P. Kloosterman & F. Lester (Eds.), *The 1990 through 2000 mathematics assessment of NAEP: Results and interpretations* (pp. 105-143). Reston, VA: National Council of Teachers of Mathematics.

Stafylidou, S., & Vosniadou, S. (2004). The development of students' understanding of the numerical value of fractions. *Learning & Instruction, 14,* 503-518.

Yeap, B.H., & Kaur, B. (2001). Semantic Characteristics that make arithmetic word problems difficult. In Bobis, J., Perry, B., & Mitchelmore, M. (Eds), *Numeracy and beyond* (Proceedings of the 24th annual conference of the Mathematics Education Research Group of Australasia). Sdyney: MERGA. Retrieved June 3, 2011 from http://www.merga.net.au/publications/counter.php?pub=pub_conf&id=1020

Chapter 3

Characteristics of Low Attainers: Behaviours, Affects and Home Backgrounds

FOONG Pui Yee   Masura GHANI

CHANG Suo Hui

This chapter profiles 375 Primary 4 low attainers in primary mathematics by investigating their characteristics in three non-cognitive domains viz-a-viz classroom behaviours, affects in learning, and home backgrounds and support. Data collected using questionnaires, interviews and student journals show that majority of them were either passive inattentive or active disruptive during mathematics lessons and often not on-task and unable to complete their classwork on time. More than half of them had low self-concept and believed that they were not good in mathematics. However, most felt that they could do better if they worked hard, paid attention in class and revised their work. Almost four tenths of the pupils did not have a study table at home. Although majority of the pupils had some form of home support to do their school work, it was mainly provided by their parents and siblings, who themselves had either primary or secondary school education.

1 Introduction

Poor performance of pupils in mathematics assessment such as class tests and national examinations is a concern of stake-holders in education. Teachers and parents often wonder why while most pupils in the same class and taught by the same teacher are able to learn mathematics, there will be some pupils who have difficulty in grasping and applying basic concepts to solve problems. There would be pupils who may fail consistently in their mathematics tests, some of whom since lower primary. Research to find possible causes of poor performance in mathematics have led to identification of categories of children been labelled with terms such as "slow learners", "low attainers", or "underachievers". The term "low attainer" has been considered as a more appropriate label for such learners as it describes the observable poor performance of the pupil through his or her grades in assessment, as opposed to "slow learners" which may imply that a pupil in an ordinary class is slower to learn than his peers due causes of some learning disabilities (Denvir, Stolz, & Brown, 1982). Hence low attainers in mathematics are primarily those pupils enrolled in the standard school mathematics curriculum who fall into the bottom 20-30 per cent of mathematical attainment in their age group in formal assessment. For the purpose of discussion in this chapter, the term "low attainers" in mathematics will be used to describe pupils in the primary schools who failed frequently in the subject.

Teachers often experience great pressure in finding the time to focus on the low attaining pupils who would normally need more individual attention from their teachers. Much of the time teachers would consider the problem of low attainment merely in terms of cognitive deficiencies using diagnostic tests to find out what errors and misconceptions the pupils may have in learning the topics that are being taught. Very often the low attainers display many gaps in their learning that teachers do not know where to begin to help them. Researchers in the field of mathematics education have realised that in addition to cognition, non-cognitive factors are equally important in the teaching and learning of mathematics (McLeod, 1992; Scheonfeld, 1989; Crane, 1996). The various TIMSS studies (e.g. TIMSS 2007) place importance in gathering detailed data of pupils' home backgrounds and resources for learning, homework, their attitudes toward mathematics and their self-confidence in learning. These data are constituents of significant non-cognitive factors that are essential in interpreting and comparing the mathematics achievement results of pupils across participating countries.

The purpose of the Low Attainers in Primary Mathematics (LAPM) study is to examine the characteristics of a cohort of Primary 4 pupils across nine primary schools in Singapore who have been identified as low attainers in mathematics. The teachers in this present study selected pupils who had failed frequently in mathematics based on results of the pupils' past years' continual assessments, semestral examinations and class tests. The main objective of this chapter is to profile these low attaining pupils' characteristics in the non-cognitive domains of learning with respect to their observed physical behaviours in the mathematics classroom, their affects in learning and their home backgrounds. The findings in this study can help teachers to be more sensitive to the affective characteristics of the low attainers. It will then enable teachers to design more responsive approaches and content for remediation with these learners.

2 Review of Literature

Research on learning difficulties to understand pupils' poor performance in mathematics has been much focused on the cognitive aspects of mathematical curricular and instruction which might be over-represented resulting in lesser studies of non-cognitive factors. Nevertheless there have been studies to link pupil achievement in mathematics with factors in the non-cognitive domains such as pupil behaviours in classroom, their affects in learning and home backgrounds.

2.1 *Pupils' behaviours in mathematics classroom*

In an attempt to seek common characteristics of low attainers with respect to their behaviours in class while learning mathematics, Foong (1999) asked each of the 24 primary school teachers involved in her study to focus attention and make observation of one such pupil under his or her charge. The teachers reported diverse characters across the 24 low attainers from various classes and levels from Primary 2 to Primary 6. Their reported observations of the low attainers' behaviours (Figure 1) were manifestations of underlying problems in the pupils' negative affective and environmental domains. Similar observations of low attaining pupils were made by teachers in other studies (Haylock, 1991; Van Kraayenoord & Elkins, 2004).

It is reasonable to assume that the experiences pupils have in mathematics classroom will be a major factor in the formation of their attitudes toward learning of the subject. At the same time, the physical behaviours such as attentiveness or disruptions that pupils exhibit maybe the outcomes of how they are feeling in the learning environment. It is arguable that behaviours and attitudes could be related since just as attitude is learned, so is the correspondent response. Hence the behaviour for a given attitude will differ for each individual. This is evident in the Haylock (1991) and Foong (1999) studies where a diverse range of behaviours have been observed illustrating that the underachievers are not a homogenous group characteristics of their classroom behaviours, albeit most of these behaviours are considered to be hindrance to learning.

- Pupil is dependent on teacher most of the time
- Pupil is passive/easily bored/not interested
- Pupil is hyperactive
- Pupil seems tired/sleepy/dreamy most of the time
- Pupil has short attention span/easily distracted
- Pupil is forgetful
- Pupil has untidy hand-writing/messy work
- Pupil often does not hand in work/homework
- Pupil has low self-esteem/confidence
- Pupil shows high level of mathematics anxiety
- Pupil has emotional problem relating to home
- Lack home support/supervision
- Excessive parental pressure/tuition
- Pupil is frequently absent

Figure 1. Observed behaviours of underachievers in Foong (1999)

2.2 *Affects in learning*

Among educators many will not disagree that favourable attitude relates to more and better learning. Hence in the Singapore Mathematics Curriculum "pentagon" framework (Ministry of Education, 2006), *Attitude* is positioned as one of the five inter-related factors to learners' success and attainment of mathematics. Primary pupils' perception of mathematics is formed largely by what happens in mathematics classroom. It is a common notion that of all the subjects taught at school, mathematics provokes the strongest emotions of dislike, anxiety and low self-concept in those students who do not seem to achieve some level of success in it. Some pupils find mathematics difficult and boring, particularly if they are having problems understanding some of the concepts which can be abstract to them. It is not surprising that many pupils do not like mathematics when they get their sums wrong or they cannot understand a word problem to start with. Hence it can be said that usually it is learning that influences such attitudes in pupils. An attitude is the tendency to respond favourably or unfavourably to a given situation or object. Many researchers have investigated the relationship between the affective domain and achievement in mathematics. In these researches, generally attitudes are often used to describe a wide range of affective responses such as perception of the subject and its value, pupils' feelings toward learning mathematics, their self-confidence or self-concept as well as motivation to learn (McLeod, 1992).

To gain understanding about grade 4 students' general attitudes towards mathematics TIMSS 2007 (Mullis, Martin, & Foy, 2008) created an index of students' positive affect towards mathematics based on students' responses to three statements: *i) I enjoy learning mathematics, ii) Mathematics is boring and iii) I like mathematics.* Across countries it found that grade 4 students at the high level of the Index of Positive Affect Towards Mathematics had higher average mathematics achievement than students at the medium or low level of achievement. To further investigate how students thought about their abilities in mathematics, TIMSS created an Index of Students' Self-Confidence in Learning Mathematics, based on students' responses to four statements about their mathematics ability: *i) I usually do well in mathematics,*

ii) Mathematics is harder for me than for many of my classmates, iii) I am just not good at mathematics and iv) I learn things quickly in mathematics. The results showed that there was a positive relationship between level of self-confidence in learning mathematics and mathematics achievement at the fourth grade. Students at the highest index of self-confidence index were found in the high mathematics achievement level, next highest among students at the medium level, and the lowest among those at the low level.

2.3 *Home backgrounds*

While school experiences can influence attitudes, pupils can also bring to class mathematical experiences that have arisen from their interaction with adults and siblings in the home environment. Hence their perceptions and attitudes toward mathematical learning could have been socially moulded beyond the teaching of mathematics in the classroom. Home environment has been found to impact significantly on mathematics test scores and the effect is more pronounced when the children are younger (Crane, 1996). Cheam and Chua (2009) recognise that home environment is one of the important factors to be considered when planning intervention programmes to help pupils at risk of mathematics difficulties. They stress that the value of parental support in promoting positive learning outcomes in the child should not be underestimated. Leone and Richards (1989) have found that pupils are most attentive to homework when completing it with a parent rather than on their own.

In a local study Leong (1999) investigated the type of parental involvement and its relationship with the academic performance of 202 Primary 5 and Primary 6 pupils from EM2 stream in a government school in Singapore. It was found that many parents could not provide direct assistance to their children in the completion of their school work. The parents tried to compensate for this with the employment of tutors for their children. There was no clear evidence to suggest that the higher the degree of parental involvement, the better the student would perform academically. It was also evident that the educational level of the pupil's parents would affect his / her academic performance. In another local study Ng (2003) investigated the possible factors relating to poor performance of the underachievers in Primary 6; it showed an association between the types of dwelling, fathers' and mothers' educational levels and underachievement. In the Wong & Quek (2009) study to explain the performance of mathematically weak pupils they found it necessary to examine how these pupils study mathematics in out-of-school context and their home environments. Spending time out of school studying or doing homework in mathematics was a significant way for pupils to consolidate their classroom learning. TIMSS 2007 constructed an Index of Out-of-School Study Time that assigned pupils to a high, medium or low level based on the amount of time they reported studying the subject. On the average internationally, pupils at the low level of the index also had lower mathematics achievement than their classmates who reported more out-of-school study time. For other home background factors which could relate to pupils' mathematics performance in school, the TIMSS study had constructed some central variables, namely i) levels of parents education; ii) speaking the language of test at home; iii) having educational resources such as books, computer and Internet connection at home, and others. According to TIMSS the information on such factors was very important in interpreting the mathematical performance of the students.

3 The Study and Conceptual Framework

In this part of the LAPM study, we attempt to profile the low attainers by investigating their characteristics in the three non-cognitive domains of i) classroom behaviours, ii) affects in learning and iii) home backgrounds, depicted in Figure 2. As supported in the review of literature, in addition to studies of pupils' cognitive development in mathematical concepts and problem solving, studies of pupils' characteristics in the non-cognitive domains are equally important for they can complete a more holistic and ecological context to interpret pupils' achievement in the subject. In this study the domain of classroom behaviours as observed by the teachers will include pupil interactions in class, their on-task behaviours and attendance. Pupil classroom interactions consist of whether they are attentive or easily distracted, good listeners, help seekers and proficiency in the English language. On-task behaviours will include pupils' engagement during individual seatwork or groupwork, completion of the classwork, and putting their best effort in the classwork. Attendance is to check for absenteeism. Domain of affects in learning will consist of pupils' perception of mathematics, their perceived value of the subject, feeling towards mathematics learning, their self-confidence, self-concept, and motivation to learn. The last domain, pupils' home backgrounds will examine pupils' demographics of gender, race, language spoken at home and types of residence; parents' educational qualifications, homework, home support in mathematics and educational resources.

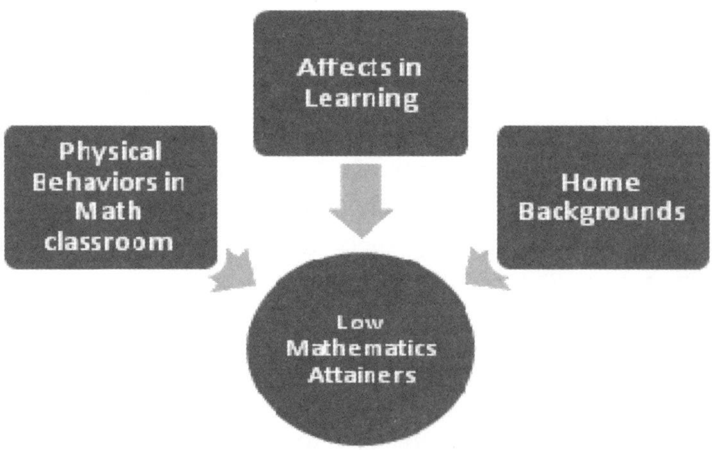

Figure 2. Characterising the low mathematics attainers in three non-cognitive domains

Hence, in this chapter we draw on the data of the study and answer the following research question:

What were the characteristics of low attaining in mathematics pupils related to school (behaviour during class, absenteeism, interacting with peers), home (support, resources) and self?

The three sub-questions addressed in this chapter are:

i) What were the characteristics of low mathematics attainers' behaviours in the mathematics classroom as observed by their teachers w.r.t pupil interactions, on-task behaviours, and attendance?
ii) What were the mathematics low attainers' affects in learning mathematics with respect to their perception of mathematics, their perceived value of mathematics, their feelings towards mathematics learning, their self-confidence, self-concept and motivation to learn?
iii) What are the characteristics of the home backgrounds of the low attainers in mathematics w.r.t their demographic profiles, parental backgrounds and support, educational resources, time spent on homework and help with homework?

4 The Subjects

A total of 375 Primary 4 pupils from nine primary schools geographically located in the north, west and east of Singapore participated in the study. The teachers who taught the low attainers were also participants of the study.

5 Procedures and Instruments

Over a period of nine months, the research team visited the nine schools to collect data on pupils. This was done in two phases. During the first phase, which was at the start of the study, the help of the teachers was sought to collect data on the profiles of the pupils using the Pupil Profile Questionnaire. Pupils were also given a Mathematics Journal each. Teachers encouraged their pupils to reflect and write about their feelings related to the learning mathematics in their journals during mathematics lessons. During the second phase the research team went to the school to meet the pupils in the study. A half day programme (about 4 hours) was organised during which the researchers administered the pupils the Pupil Maths Learning Questionnaire and interviewed them in groups of four or five about their affects in mathematics learning. The interviews were audio-recorded. During the half day programme, pupils first completed the questionnaire with the help of researchers who were at hand to explain any word or item the pupils did not comprehend. Next the pupils were scheduled for the interviews. The time pupils had between their turns for the interview was spend playing in-door games which were organised by the research team and some volunteers (pre-service teachers). At the end of the school visit the researchers sought the help of the teachers to complete the Pupil Behaviour Questionnaire. Teachers were given a questionnaire for every pupil in the study and they completed them in the following weeks. The completed questionnaires were posted by the schools to the researchers.

5.1 *Pupil profile questionnaire*

The teachers gave individual pupils the Pupil Profile Questionnaire (see Appendix E) to take home for their parents to fill out the required information and collected them for the researchers. The information required were the pupils' demographic profiles, their parental educational backgrounds and support, educational resources at home, time spent on homework and help in homework. In this chapter we will present data on the following items:

- Gender
- Race
- Language(s) spoken at home
- Type of dwelling
- Parents' educational background
- Number of hours the pupils spend on homework
- Support and resources available for the pupils to do homework

5.2 *Pupil behaviour questionnaire*

Each teacher who was involved in teaching the low attainers in the study was given the Pupil Behaviour Questionnaire (see Appendix F) to respond on each of their pupil's behaviours such as classroom interactions, on-task behaviours, work attitudes and attendance. In this chapter we will explore the teachers' responses on the following items:

- Is the pupil attentive when you are teaching?
- Is the pupil on task during groupwork?
- Is the pupil on task during individual seatwork?
- Does the pupil seek help from the teacher when in doubt?
- Does the pupil complete his / her classwork on time?
- Does the pupil give his / her best when doing work in class?
- Is the pupil a good listener?
- Is the pupil easily distracted?
- Is the pupil often away from school without valid reasons?
- How is the pupil's proficiency in English Language?

The teachers' response to an open-ended question asked will also be presented to better understand the pupil's mathematics attainment.

5.3 *Pupil maths learning questionnaire*

The researchers gave every pupil the Pupil Maths Learning Questionnaire (see Appendix G) to complete on the day of the pupil interviews. Pupils responded to statements relating to three dimensions of affects: i) their perceptions towards mathematics, ii) their perceived value of mathematics, and iii) their self-confidence. In addition, they were asked to respond to statements relating to math work and help at home. The pupils were given sufficient time to complete the questionnaire and if they were unsure of the statements the researchers were available for clarification. Specific to this chapter, we will only explore pupil responses to the following items:

- Do your parents check your maths work?
- Who help you at home when you have difficulties in your maths homework?
- Maths is easy.
- Maths is important.
- Maths is fun.
- Maths is useful in daily life.
- I am good at maths.
- I give up when a problem looks difficult.
- I get stuck when solving maths problem.

5.4 *Pupil interview prompts*

After the pupils completed the Pupil Maths Learning Questionnaire, they were interviewed in groups of four or five pupils by the researchers. The prompts used for the interviews belonged to three categories, mathematics lessons, homework and self (see Appendix H). In this chapter we only focus on the prompts related to self. The following prompts were used to engage pupils in talking about their self-beliefs in mathematics:

Prompt 1: Do you think you are good or not good in maths? How can you tell?

Prompt 2: Do you think you can do better in maths? Why or why not?

5.5 *Pupil journal*

Pupils were asked to write in their journals about "How do you feel about learning mathematics?" (see Appendix I). They were to express themselves in words and to draw a picture to show their feelings and attitudes.

6 Data and Findings

The data collected using the Pupil Profile Questionnaire, Pupil Behaviour Questionnaire, Pupil Maths Learning Questionnaire, pupil interviews and pupils' journal writings were analysed and presented hereof in order of the three sub-questions and non-cognitive domains:

- Classroom behaviours
- Affects in learning mathematics
- Home background and support

6.1 *Classroom behaviours*

In the Pupil Behaviour Questionnaire (Appendix F), the teachers responded to various statements about individual pupils' behaviour in the classroom. A total of 371 pupil questionnaires were completed by the teacher participants. The various behaviours are categorised under three types: *classroom interactions, on-task behaviours and attendance.*

6.1.1 Classroom interactions

Only about 30% of the low attainers were attentive while the teachers were teaching and the rest were sometimes, seldom or never paid attention as shown in Table 1. This was verified when the teachers noted that more than 60% or the pupils were easily distracted as shown in Table 2.

Table 1
Teachers' response to "Is the pupil attentive when you are teaching?"

	Frequency n (%)			
	Always	Sometimes	Seldom	Never
Attentive	111 (29.9)	214 (57.7)	45 (12.1)	1 (0.3)

Table 2
Teachers' response to "Is the pupil easily distracted?"

	Frequency n (%)	
	Yes	No
Easily distracted	225 (60.6)	146 (39.4)

From these observations, majority of the low attainers were not always attentive and they were easily distracted. Figure 3 shows some of the teachers' comments on pupils' behaviours when they were asked to give further information to help understand these pupils' mathematical attainment. There were teachers who attributed these low attainers' behavioural problems to their attitudes and their poor performance in mathematics: "*Have behavioural problems that translate in to underperformance in academic*" and "*Severe lack of concentration and inability to achieve stem from extreme bad attitudes.*"

On the active disruptive learners	On the passive inattentive pupils
▪ *Playful/ difficulty paying attention.* ▪ *Often caught talking, drawing or doing other activities when teacher is teaching.* ▪ *Easily distracted by himself with his toys or belongings.* ▪ *Always fidgeting, playing with stationary, talking with classmates or moving from seat to seat.* ▪ *Is more attentive in class when he is able to understand the concepts.* ▪ *Not able to sit and listen for a long period of time.* ▪ *Seeks attention and not focused.* ▪ *Very expressive, not afraid to answer questions but can be disruptive during lessons.* ▪ *His learning style is more towards visual and kinesthetic.*	▪ *Dream in class/ tends to dream.* ▪ *Daydreams and need reminder to stay focus/drifts off during lesson.* ▪ *Very inattentive, in a world of his own most of the time.* ▪ *Too timid/ keeps to herself/very quiet.* ▪ *Seldom participate in class discussion/ does not interact.* ▪ *Lack of concentration in class but when manipulatives are used, he cooperates well. He is more of a kinesthetic learner.*

Figure 3. Teachers' description of low attainers' behaviours in mathematics classroom

However according to the teachers, as shown in Table 3, 63.3% of the low attainers were good listeners when they were called upon in class and about 78% of them, as shown in Table 4, have average level of English Language proficiency. Only about 11% had good English Language proficiency.

Table 3
Teachers' response to "Is the pupil a good listener?"

	Frequency n (%)	
	Yes	No
Good listener	235 (63.3)	136 (36.7)

Table 4
Teachers' response to "How is the pupil's proficiency in English Language?"

	Frequency n (%)		
	Good	Average	Poor
English Language proficiency	42 (11.3)	289 (77.9)	40 (10.8)

Table 5 shows that only about 15% of the low attainers always sought the help of the teachers when in doubt. More than 44% seldom or never sought the teachers' help when they did not understand.

Table 5
Teachers' response to "Does the pupil seek help from the teacher when in doubt?"

	Frequency n (%)			
	Always	Sometimes	Seldom	Never
Seek help	57 (15.4)	148 (39.9)	122 (32.9)	44 (11.9)

In the open response question, the following elaborations by the teachers (Figure 4) on pupils' passive behaviours could be at times explained by lack of language proficiency of the pupils:

- *Very quiet in class, difficult to gauge how much he understands.*
- *Shy who does not seek help when in doubt.*
- *Well behaved but quiet boy who does not seek help when in doubt.*
- *Not totally focused, inability to comprehend questions.*
- *Does not answer question when asked, seem to have difficulty comprehending, unable to follow instructions.*
- *Very poor command of English Language.*
- *Has problem reading.*
- *Unable to follow lesson due to extremely low proficiency in English Language.*
- *More comfortable in speaking Mandarin.*

Figure 4. Teachers' descriptions of low attainers' behaviours in help seeking and language proficiency

6.1.2 On-task behaviours

Table 6 shows the teachers' responses on the pupils' on-task behaviours during seatwork and group work. For group work only a third of the low attainers were always on task and slightly more of them about 36% were more focused during individual seatwork. Nevertheless majority were sometimes on-task. With regard to whether these pupils were able to complete their classwork on time, only 39.6% were able to while the rest, 47.2% sometimes and 10.8% seldom complete the work given in class. Nine pupils never completed their work on time. Similar lower proportion of the low attainers gave their best when doing work in class while the majority was sometimes, seldom or never gave their best at work.

Table 6
Teachers' response to pupil on-task behaviours

	Frequency n (%)			
On-task statements	Always	Sometimes	Seldom	Never
Is the pupil on task during groupwork?	109 (29.4)	230 (62.0)	31 (8.4)	1 (0.3)
Is the pupil on task during individual seatwork?	134 (36.1)	199 (53.6)	34 (9.7)	2 (0.5)
Does the pupil complete his/her classwork on time?	147 (39.6)	175 (47.2)	40 (10.8)	9 (2.4)
Does the pupil give his/her best when doing work in class?	137 (36.9)	189 (50.9)	42 (11.3)	3 (0.8)

Some teachers attempted to give reasons for the low attainers' inability to complete their work because of weakness in mathematical understanding and difficulty in following instructions. On a positive note, some teachers observed that pupils could complete their tasks with teacher supervision and more effort from the pupils. The teachers substantiated their observations with the following descriptions in Figure 5.

Negative on-task behaviours	Positive on-task behaviours
▪ *Tends to stop work after 1 or 2 steps. He rarely finishes his sums.* ▪ *Unable to apply what she has learnt. Forget easily.* ▪ *Tend not to follow instructions unless checked by the teacher.* ▪ *Very careless and clueless in his work.* ▪ *Has difficulty in grasping simple math concepts e.g. division.* ▪ *Tend to forget the concept, need help to recall the facts.* ▪ *Relies quite a bit on partner for guidance* ▪ *Able to understand concepts but poor retention.* ▪ *Prefers to talk and wait for teacher to give him answers.* ▪ *Unable to perform even in small group of 1 teacher to 5 pupils.*	▪ *Has made improvement in term 3 in his concentration and quality of work.* ▪ *Can do work with teacher's supervision.* ▪ *Requires lots of concentration to absorb explanation and instruction, will be able to improve with hard work.* ▪ *Slow in absorbing but will be able to catch up if she works harder.* ▪ *Sometimes careless but given the opportunity, she is capable of scoring good grades.*

Figure 5. Teachers' descriptions of low attainers' on-task behaviours

6.1.3 Attendance

School attendance of the low attainers was high with 91.6%, except 31 pupils who were often absent without valid reasons. For these pupils the teachers had the following comments in Figure 6.

- *Gets sick easily and stays away from school often.*
- *Often away from school as he is often sick.*
- *Very weak in maths and frequently absent due to asthma.*
- *Irregular attendance due to asthma.*
- *Very frequently absent hence unable to catch up. Also not motivated to acquire skills.*
- *Usually absent between 1 to 3 days a week.*
- *Misses lessons quite often but still able to catch up to some extent.*
- *Irregular attendance and often comes to school late, not motivated to learn at all.*

Figure 6. Teachers' comments on low attainers' absence from school

6.1.4 Learning disability (Special needs)

It is revealing that the teachers have identified in some of the low attainers some learning disability due to neurological or emotional problems. Some of these pupils were officially diagnosed while in some cases the teachers suspected these pupils might have some special needs. Some of their observations are shown in Figure 7.

- *Child is diagnosed as having development disorder (official diagnosis from EP).*
- *Has mild dyslexia.*
- *Diagnosed as having learning difficulties, has referred child to META. Not able to do basic math skills.*
- *Has been certified by EP as having low IQ has difficulty staying focused, unable to finish exam within stipulated time.*
- *Suspected to have ADHD or attention problem.*
- *Seems to be in special needs, but not quite supported/assessed.*
- *Took more than a year to familiarize with me before she started talking to me. She never looks someone in the eyes while talking.*
- *Brings a lot of emotions to the class. She is frequently upset over small issues. She at times does not respond despite been called many times. This situation is also the same at home.*
- *Child is recently undergoing emotional difficulties due to the loss of his mother. Maths results have shown a dip lately.*

Figure 7. Teachers' observations of some pupils with special needs

6.2 *Affects in learning mathematics*

Domain of affects in learning will consist of pupils' perception of mathematics, their perceived value of the subject, feeling towards mathematics learning, their self-confidence, self-concept, and motivation to learn. Data were collected using the Pupil Maths Learning Questionnaire, interviews and journal writing. Discussions of the various results attempted to link and integrate these various constructs of pupils' affects in learning to their mathematics achievement and observed classroom behaviours by the teachers.

6.2.1 Pupils' perception of mathematics

Pupils' perception of mathematics was gauged by two statements from the Pupil Math Learning Questionnaire shown in Table 7.

Table 7
Pupils' perception of mathematics

	Frequency n (%)		
Statement	Agree	Not sure	Disagree
Maths is easy	110 (30.2)	210 (57.7)	44 (12.1)
Maths is fun	242 (66.5)	94 (25.8)	28 (7.7)

About a third of the pupils perceived mathematics to be easy while 57.7% were not sure and 12% disagree with this statement. However a large majority, 66.5% of low attainers perceived mathematics to be fun, with 25.8% not sure if it so and 7.7% did not think it is fun. When we combined the two statements for an index of pupils' perception of the subject, shown in Table 8, it would seem that a mean of 2.38 would indicate that these pupils' perception was relatively ambivalent toward the unsure category. A score of 3 would mean the pupils have very positive perception of the subject and score of less than 2 would mean a negative perception of the subject.

Table 8
Mean of pupils' perception of mathematics

Statement	Sum	Mean
Maths is easy	756	2.1850
Maths is fun	896	2.5896
Total	1652	2.3873

6.2.2 Pupils' perceived importance of mathematics

88.5% and 65% of the pupils perceived that mathematics is important and it is useful in daily life respectively. The very high number of pupils who perceived mathematics is important could be due to their school experience as mathematics is compulsory and one of the main examination subjects for Primary School Leaving Examination. Only one pupil disagreed that it is important. About 35% of the pupils were unsure or disagree with the statement that mathematics is useful in daily life, possibly of the limited exposure in the classroom where learning of mathematics was focused more on abstraction than link to real life experiences for pupils' awareness. On the whole, the pupils had a positive perception of the value of mathematics based the mean index of 2.7 for the two statement on a scale of 3 to 1.

Table 9
Pupils' perceived importance of Mathematics

	Frequency n (%)		
Statement	Agree	Not sure	Disagree
Maths is important	322 (88.4)	42 (11.3)	1 (0.3)
Maths is useful in daily life	238 (65.4)	96 (26.4)	30 (8.2)

Table 10
Mean of pupils' perceived importance of Mathematics

Statement	Sum	Mean
Maths is important	999	2.8873
Maths is useful in daily life	892	2.5780
Total	1891	2.7327

6.2.3 Pupils' self-confidence in mathematics

Pupils' self-confidence in mathematics was gauged by the three statements in the Pupil Maths Learning Questionnaire shown in Table 11. Only about 10% of the low attainers thought themselves to be good in mathematics, 53% thought they were sometimes good and sometimes not good, and 34% thought they were never good at it. In the same trend, about 11% thought they never got stuck when solving mathematics, 56% got stuck sometimes and 33% got stuck always. About half the pupils (49.7%) would give up sometimes when the problem looked difficult while 36.4% never gave up when the problem looks difficult. Only 50 pupils or 13.7% would always give up when the problem looks difficult. Combining the three items to give an index of mean 2.0 on a scale of 1 of low confidence to a scale of 3 for high confidence, this would indicate that the low attainers' confidence was unstable, sometimes they were confident and sometimes they were not.

Table 11
Pupils' self-confidence in mathematics

	Frequency n (%)		
Statement	Always	Sometimes	Never
I am good at maths	37 (10.2)	287 (78.9)	40 (10.9)
I give up when a problem looks difficult	50 (13.7)	181 (49.7)	133 (36.6)
I get stuck when solving maths problems	120 (33.0)	204 (56.0)	40 (11.0)

Table 12
Mean of pupils' self confidence in mathematics

Statement	Sum	Mean
I am good in Math	725	1.99
I give up when a problem seems difficult	811	2.23
I get stuck when solving mathematics problems	648	1.78
Total	747.25	2.00

6.2.4 Pupils' self-concept and motivation in mathematics

To probe further their confidence about themselves in the learning of mathematics, these pupils were interviewed in groups of four or five on their self-concept. The interviewers asked the question *"Do you think you are good or not good in maths? How can you tell?"* The pupils' responses were broadly categorised into three types of self-concepts as shown in Table 13.

Table 13
Pupils' responses to "Do you think you are good or not good in Maths?"

Statement	Frequency n (%)
I think I am no good in maths.	187 (54.2)
I think sometimes good, sometimes bad, depends.	112 (32.5)
I think I am good in maths.	46 (13.3)

54% of the pupils thought that they are not good in mathematics. When probed on how they could tell that they were not good, ninety or half of them said that they had failed their tests and examinations. There was a strong sense of being a failure in mathematics for these young learners. Common expressions are shown in Figure 8.

- *No. When you fail, you are not good.*
- *Not good, fail maths since P3.*
- *Not good at all because I fail in maths all the time.*
- *Not good because I fail only maths and pass in all the other subjects.*
- *No, I always fail from P1, below 20.*
- *No, I obtain Band 4 for my SA1.*
- *Not good, I am scared that I will fail the exam.*
- *Not good, started failing in P4. I want to pass in math and feel good.*
- *Bad, Everytime I fail. I feel sad.*

Figure 8. Common expressions from pupils who thought they are not good in mathematics

A third of the pupils, about 32.5% were ambivalent about how they measured themselves in mathematics performance. Most felt that they were sometimes good when they could understand or if they focused and achieve good results. Other times they would feel bad when they fail tests or exams. Some of their expressions and reasons are given in Figure 9.

- *Sometimes, according to results.*
- *Sometimes. When I do difficult sums, I am no good and when I do simple sums, I am good.*
- *Don't know if I am good, I almost passed SA1.*
- *A little good, sometimes I was the only one who got the answer correct in my group.*
- *50-50, my family tells me I am good in maths, but I always make careless mistakes.*
- *Average as when the teacher teaches slowly I can understand.*
- *Only 50-50 because sometimes maths is hard, sometimes easy.*
- *Sometimes no good when I did not get a sticker.*
- *Sometimes. Maths is very hard and I make careless mistakes.*

Figure 9. Common expressions from pupils who were ambivalent about how they measured themselves in mathematics

Even though they were identified as low attainers in mathematics, 46 pupils or about 13.3% of the group had a positive self-concept about their mathematics ability. Most felt good or "OK" as they had experience past success like the recent Semestral Assessment 1 tests or passed in Primary 3; and a few attributed their positive outlook to their work attitudes (Figure 10).

- *Good, P1 is very good in mathematics as my parents help me.*
- *Good, because I listen to teachers and can solve the sums.*
- *Yes, I got 70 plus for the last exam.*
- *Good, I got full marks in my test.*
- *Yes, teacher said I am capable to get Band 2.*
- *Good. I got 39.5 and my teacher said the passing mark is 40.*
- *Good even though I fail SA1, I know I must work harder.*
- *Good. I come to school and learn all the time.*
- *Good. P1 and P2 I passed. P3 I failed because our class very noisy. P4 CA1 I failed but my teacher guided me and I passed SA1.*

Figure 10. Common expressions from pupils who had a positive self-concept about their mathematics ability

When pupils were asked *"Do you think you can do better in maths? How can you do better?"*, overwhelmingly about 90% of the low attainers believed that they can do better as shown in Table 14.

Table 14
Pupils' response to "Do you think you can do better in maths?"

Do you think you can do better in maths?	Frequency n (%)
Yes, I can.	309 (89.6)
No, I don't think so.	31 (9.0)
I do not know if I can do better.	5 (1.4)

They strongly believed that if they work harder, revise at home, practice more with assessment books, have a tutor and stay focus during mathematics lesson, they could improve their achievement. On the other hand about 31 of the low attainers did not believe that they could improve. They were defeated by past failure, dislike of mathematics and its difficulty, no one to help them and self-depreciation. Some of the pupils' comments on how they can do better in mathematics are shown in Figure 11.

Yes, I can do better in maths	No, I don't think I can be better
▪ *Doing assessment books and study everyday.* ▪ *Yes, by asking my parents to buy assessment books and revise everyday at home.* ▪ *Yes, by studying hard and listening to teacher.* ▪ *Yes, if I can remember I will do better* ▪ *Yes, study hard and not play games. Do more problem sums.* ▪ *Yes, focus during math lessons and not talk to friends.* ▪ *Yes, I can I try my best.* ▪ *Yes, doing assessment and get a home tutor.* ▪ *Yes, my parents give me tutor, then I can do better.* ▪ *Yes, must do their own revision at home or library and prepare their own study timetable without their parents forcing them.*	▪ *No, I hate maths.* ▪ *No, I get confused.* ▪ *No, I am scared, I still fail.* ▪ *No, don't think so, I fail and my parents scold me.* ▪ *No, I always make mistakes. Don't think I can do better.* ▪ *No, there is no one to help at home.* ▪ *No, there is no one to help me.* ▪ *No. Mathematics is not like English and other subjects. English we can read write the answers. Math we have to remember so many things.* ▪ *No, Math is not easy. I have tried my best already…* ▪ *No, sometimes I am too tired to listen to the teacher.* ▪ *No, because in class, all are good and I am not good.* ▪ *No, I am lazy.*

Figure 11. Pupils' comments on why they can/cannot do better in mathematics

The data from the Pupil Maths Learning Questionnaire and the interviews showed a small group of defeated pupils, about 10%, who thought that they were never good in mathematics and did not think that they could improve. Except for this group, the data also showed that even though majority of the low attaining pupils had low self-concepts in mathematics and some were unsure of themselves, they were all very positive and believed that they could do better. This positive trait was borne out in the Pupil Behaviour Questionnaire shown in Table 15.

Table 15
Teachers' response to "Is the pupil motivated to learn?"

	Frequency n (%)	
	Yes	No
Motivated	287 (77.4)	84 (22.6)

77.4% of the pupils were observed by their teachers as having motivation being eager to learn and try their best. On the other hand,

22.6% were not motivated for lack of interest or poor attitudes. Figure 12 shows the teachers' descriptions of the pupils' attitude.

Child is motivated	Child is not motivated
▪ *Tries hard and motivated to learn but has difficulty grasping concepts.* ▪ *Tries her best and is one of the better able pupils.* ▪ *Tries her best but very weak in math.* ▪ *Really tries his best and never gives up.* ▪ *Has positive attitude and eager to learn.* ▪ *Improve and getting better however he is easily influenced by his peers.* ▪ *Tries her best will be able to do well with regular practice.* ▪ *Will be able to catch up if she works harder.* ▪ *Good understanding of math concepts can go far with encouragement and motivation.* ▪ *Very motivated, has improved a lot especially in problem sums requiring the drawing of models.* ▪ *Very careless but an eager learner otherwise.* ▪ *Cheerful and eager to learn.*	▪ *No confidence in doing maths since primary.* ▪ *No interest in maths since P2.* ▪ *Does not give his best most of the time.* ▪ *Lack motivation to study, parents' divorce. He gets his ways from father.* ▪ *No motivation to learn very disorganised.* ▪ *He has an attitude problem and he finds math very difficult only because he is uninterested.* ▪ *Tends to find the easier way out of completing her work through copying rather than trying.* ▪ *Does not have perseverance when it comes to doing mathematics.* ▪ *Is able to achieve but poor attitude prevents his achievement.*

Figure 12. Teachers' description of the pupils' attitudes

6.2.5 Pupils' feeling towards learning mathematics

In this study journal writing was used as a strategy to gain insight into the low attainers' feelings toward the learning of mathematics. Their journal writing was elicited by this question: "How do I feel about learning mathematics?" and they were also required to draw a picture to illustrate how they feel. From their writings and illustrations as shown in the following, they showed that in the learning of mathematics there aroused in these low attainers a diverse range of emotions from feeling of happiness, dislike, anger, confusion, stress and anxiety to boredom as shown in Figures 13a, 13b, 13c, 13d, 13e, 13f, and 13g.

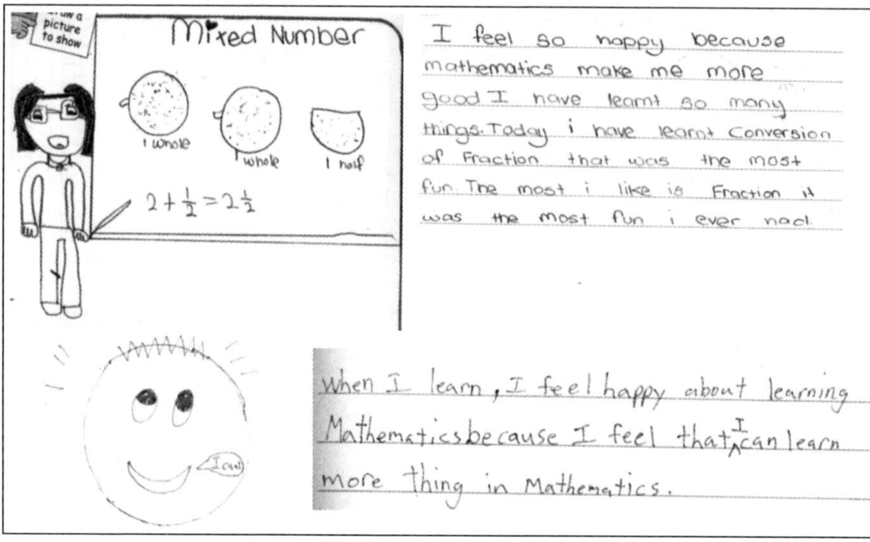

Figure 13a. I love maths…

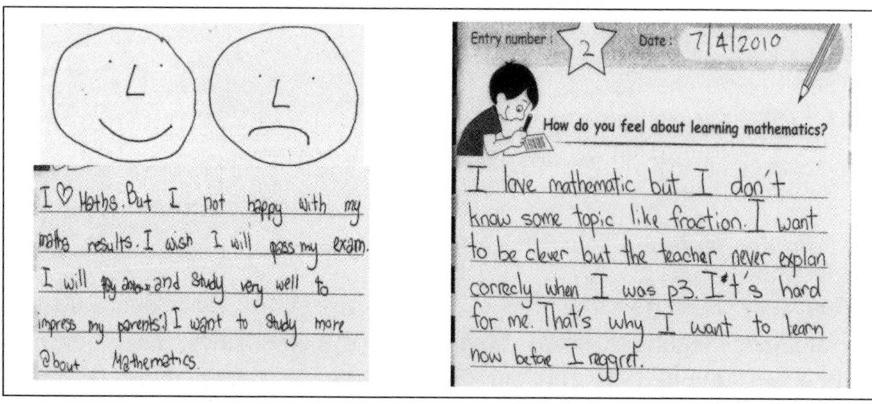

Figure 13b. I love maths but…

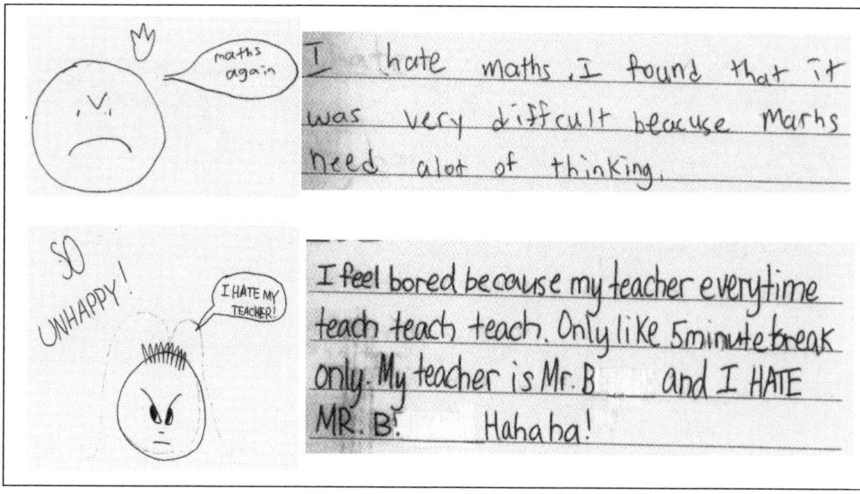

Figure 13c. I hate maths…

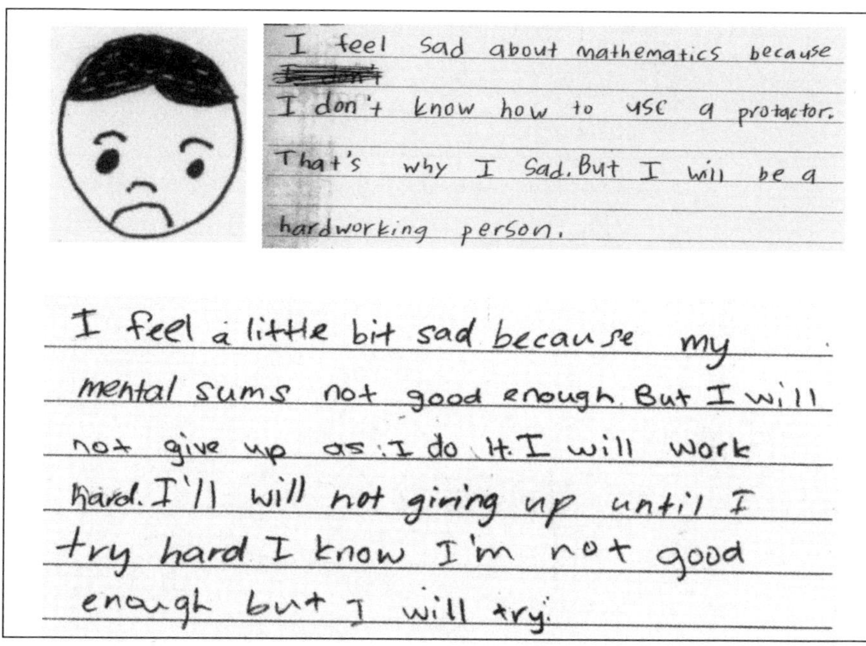

Figure 13d. I feel sad…

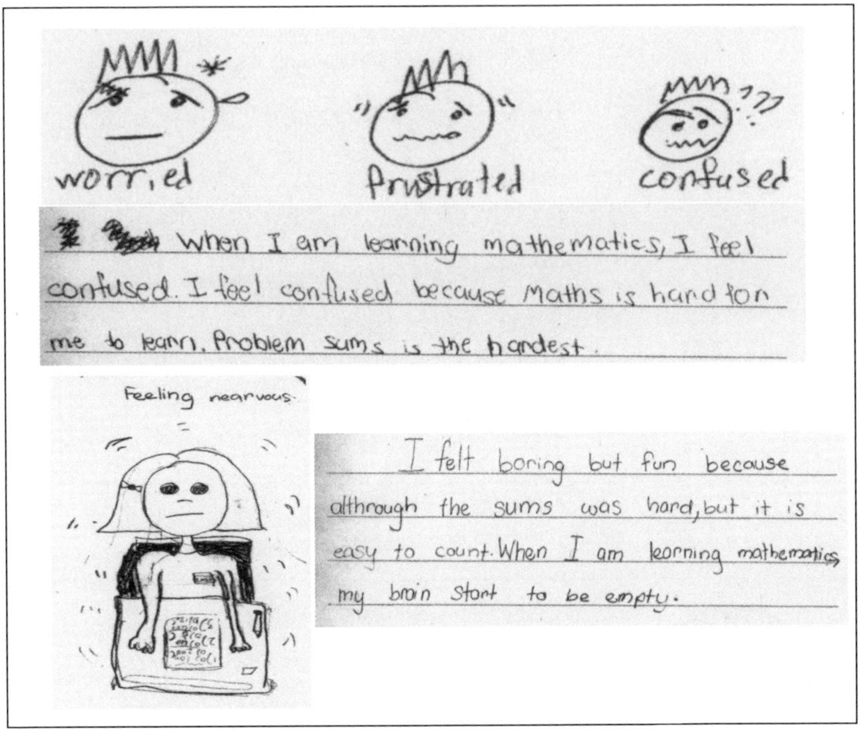

Figure 13e. I feel confused, frustrated, nervous…

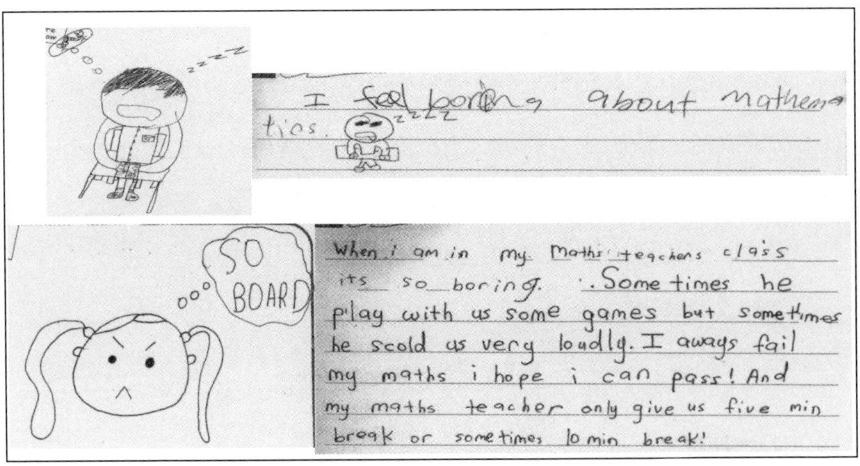

Figure 13f. I feel bored…

Figure 13g. I feel stressed…

6.3 *Home background and support*

The Pupil Profile Questionnaire (Appendix E) was used to collect data on the pupils' demographic profiles, their parents' educational qualifications, educational resources at home, time spent on homework and help in homework.

6.3.1 Pupils' demographic profile

Figure 14 shows that in the study there were more male pupils (52.8%) than female pupils (47.2%). About half the pupils were Malay with Chinese making up a third and the rest Indian and others as shown in Figure 15.

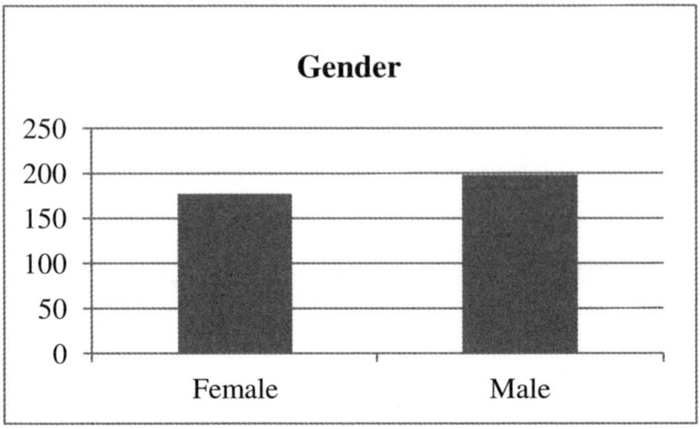

Gender	n (%)
Female	177 (47.2)
Male	198 (52.8)

Figure 14. Gender of pupils in the study

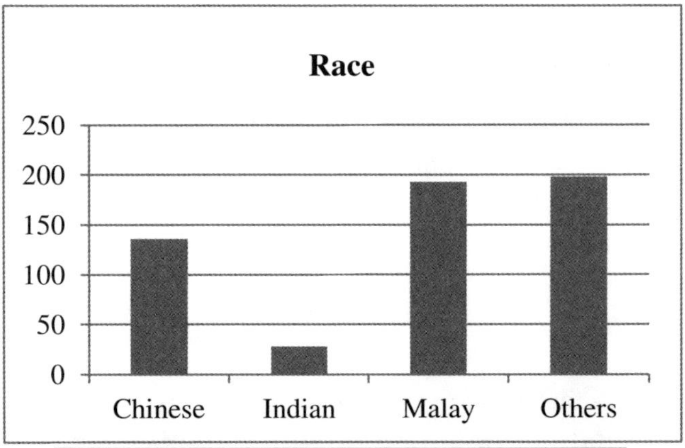

Race	n (%)
Chinese	136 (36.2)
Indian	28 (7.5)
Malay	193 (51.5)
Others	1 (4.8)

Figure 15. Race of pupils in the study

As shown in Figure 16, only 11.7% of the pupils used mainly English as a language of communication at home. In all about 27.5% of the pupils used English and their mother tongue language or Mandarin. More than 50% of the pupils spoke only their mother tongue languages or mandarin at home (32.0% Malay, 20.3% Mandarin, and 2.1% Tamil).

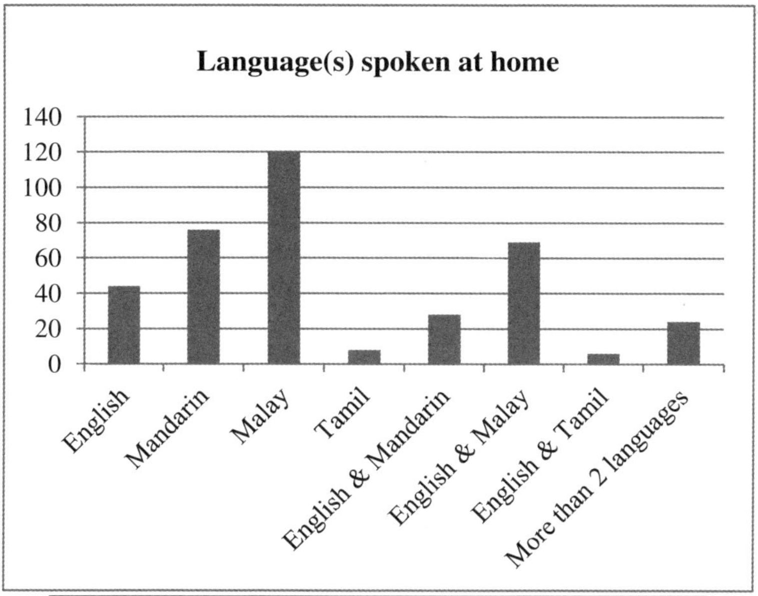

Language (s) spoken at home	n (%)
English	44 (11.7)
Mandarin	76 (20.3)
Malay	120 (32.0)
Tamil	8 (2.1)
English & Mandarin	28 (7.5)
English & Malay	69 (18.4)
English & Tamil	6 (1.6)
More than 2 languages	24 (6.4)

Figure 16. Language (s) spoken at home by pupils in the study

Majority of the pupils lived in public housing, i.e. Housing Development Board (HDB) flats. Figure 17 shows that the highest percentages of pupils, i.e. 48.2% and 23.5% lived in 4-room and 3-room HDB flats respectively.

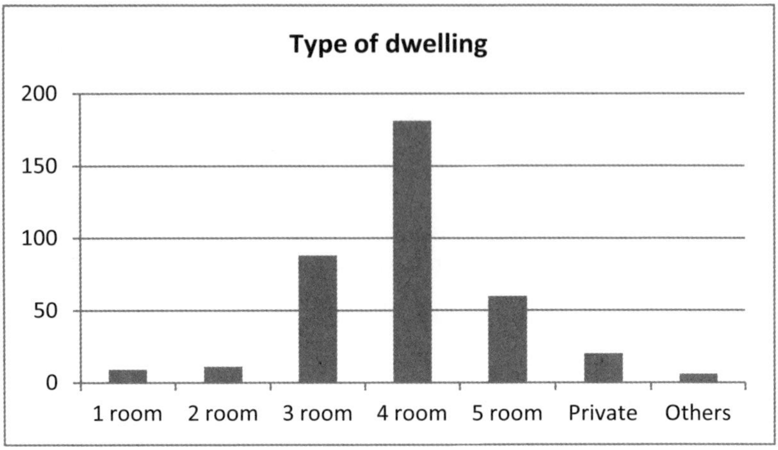

Type of dwelling	n (%)
HDB 1-room flat	9 (2.4)
HDB 2-room flat	11 (2.9)
HDB 3-room flat	88 (23.5)
HDB 4-room flat	181 (48.2)
HDB 5-room flat	60 (16)
Private	20 (5.3)
Others	6 (1.6)

Figure 17. Type of dwelling of pupils in the study

6.3.2 Parents' educational backgrounds

From Figures 18 and 19, it is apparent that about 45% and 25% of the pupils had both parents with secondary school qualification and primary school educations respectively. 16% of the pupils had fathers with VITB / ITE qualifications. Less than 9% of the parents had tertiary educational qualifications.

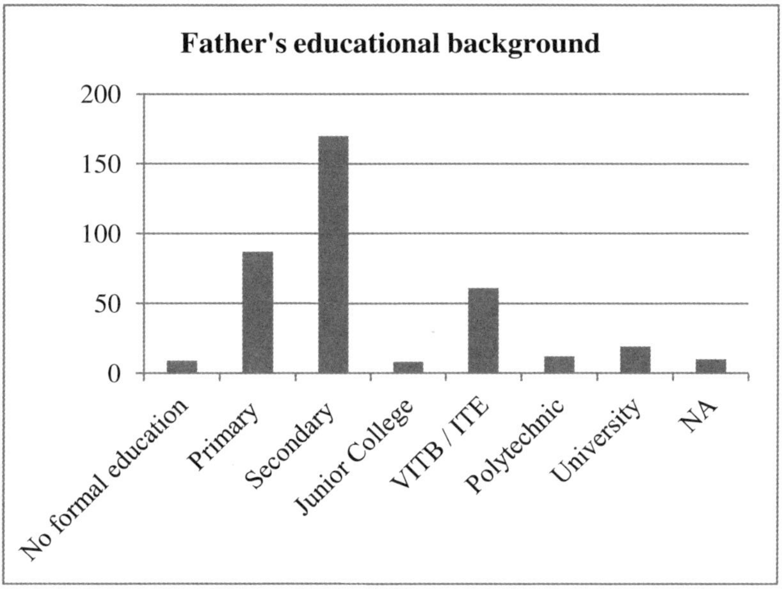

Father's educational background	n (%)
No formal education	9 (2.4)
Primary	87 (23.2)
Secondary	170 (45.3)
Junior College	8 (2.1)
VITB / ITE	61 (16.3)
Polytechnic	12 (3.2)
University	18 (4.8)
N.A.	10 (2.7)

Figure 18. Father's educational background of pupils in the study

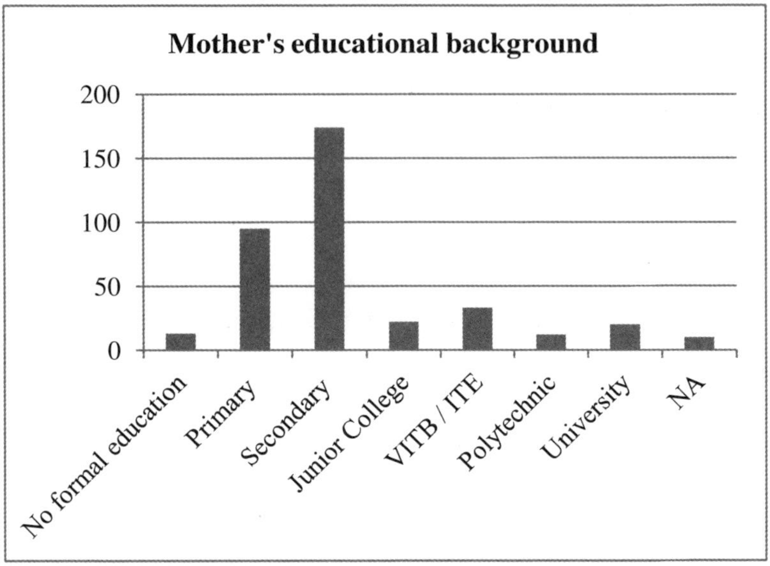

Mother's educational background	n (%)
No formal education	13 (3.5)
Primary	95 (25.3)
Secondary	174 (46.4)
Junior College	22 (5.9)
VITB / ITE	33 (8.8)
Polytechnic	12 (3.2)
University	20 (5.3)
N.A.	6 (1.6)

Figure 19. Mother's educational background of pupils in the study

6.3.3 Homework and home support

As shown in Figure 20, 68% of the pupils reported that they spent about 1 hour per day on homework while 14.1% of the pupils reported that they did not spent any time on homework. Another 14.1% of the pupils spent 2 hours on homework.

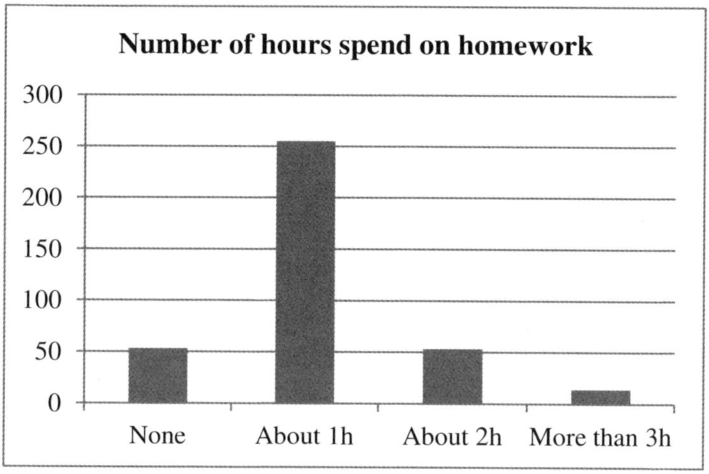

Number of hours spend on homework	n (%)
None	53 (14.1)
About 1 hour	255 (68.0)
About 2 hours	53 (14.1)
More than 3 hours	14 (3.8)

Figure 20. Number of hours pupils in the study spends on homework

As shown in Figure 21, 28.9% of the pupils reported in the Pupil Maths Learning Questionnaire that their parent(s) did not check their mathematics work at home while 36.5% reported that parent checked only whenever they were free and 15% were checked at least once a week. 19.2 % of the pupils had their work checked every day.

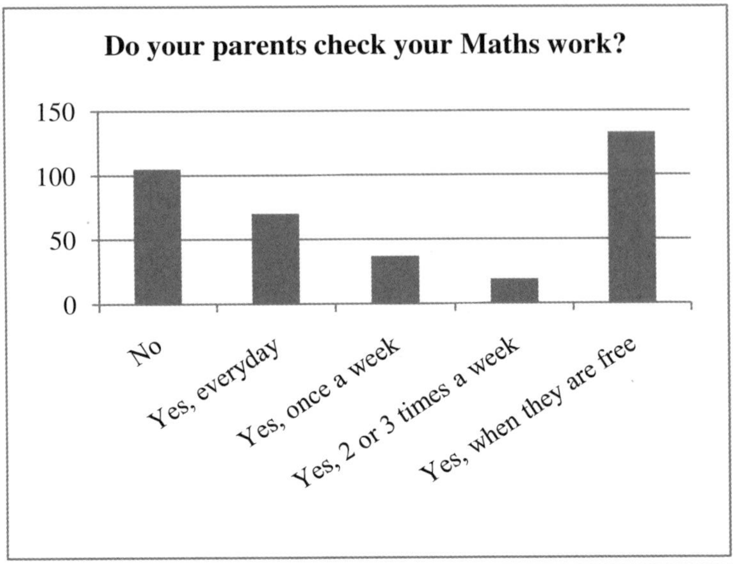

Do your parents check your Maths work?	n (%)
No	105 (28.9)
Yes, everyday	70 (19.2)
Yes, once a week	37 (10.2)
Yes, 2 or 3 times a week	19 (5.2)
Yes, when they are free	133 (36.5)

Figure 21. Pupil responses to "Do your parents check your Maths work?"

Figure 22 shows the data for who helps the pupil at home in mathematics. Pupils reported the people who help them in mathematics at home were their siblings (21.1%), their mothers (20.3%) or their fathers (11.7%). About 22% also sought help from both parent and siblings. 9% of the pupils had help from aunt, uncle or grandparent. 7.7% had no one to seek help in mathematics at home. Also, Figure 23 shows that about 40% of the pupils had some form of tuition in mathematics outside of school. There were private one-one tuition (13.6%), private group tuition (12.5%) and community-organised tuition (10.1%). From Figure 24, it is apparent that 56.3% reported that they had their own study table at home while the rest had no study table of their own.

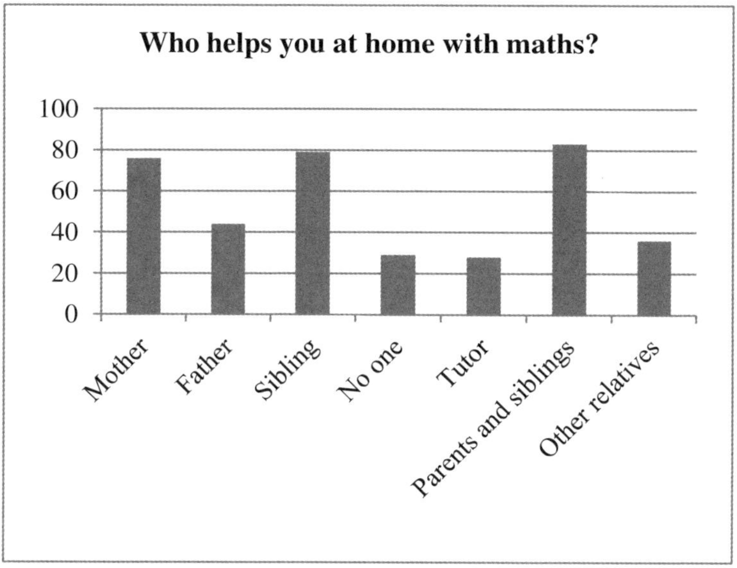

Who help you at home in mathematics?	n (%)
Mother	76 (20.3)
Father	44 (11.7)
Sibling	79 (21.1)
No one	29 (7.7)
Tutor	28 (7.5)
Parents and siblings	83 (22.1)
Other relatives	36 (9.6)

Figure 22. Pupil responses to "Who help you at home in mathematics?"

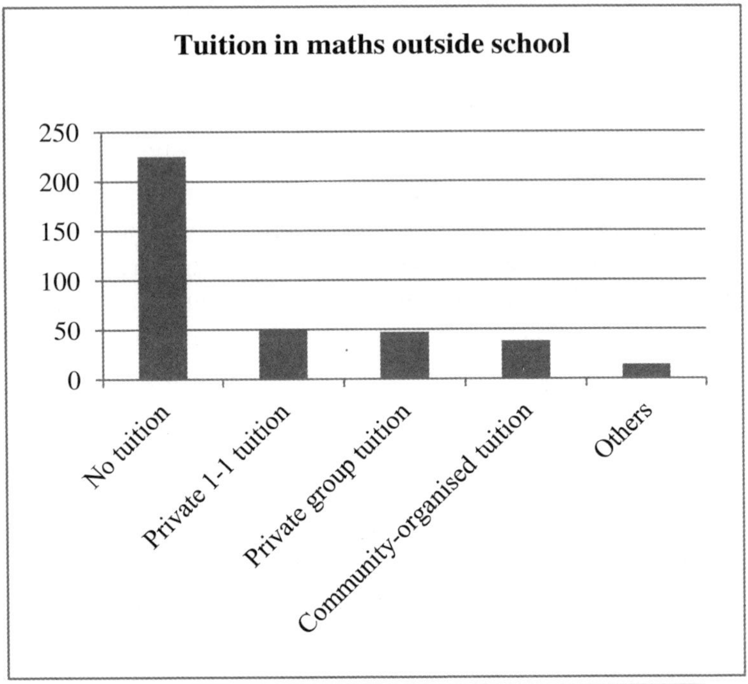

Tuition in mathematics outside school	n (%)
No tuition	225 (60.0)
Private 1-1 tuition	51 (13.6)
Private group tuition	47 (12.5)
Community-organised tuition	38 (10.1)
Others	14 (3.7)

Figure 23. Pupil responses to "Tuition in mathematics outside school"

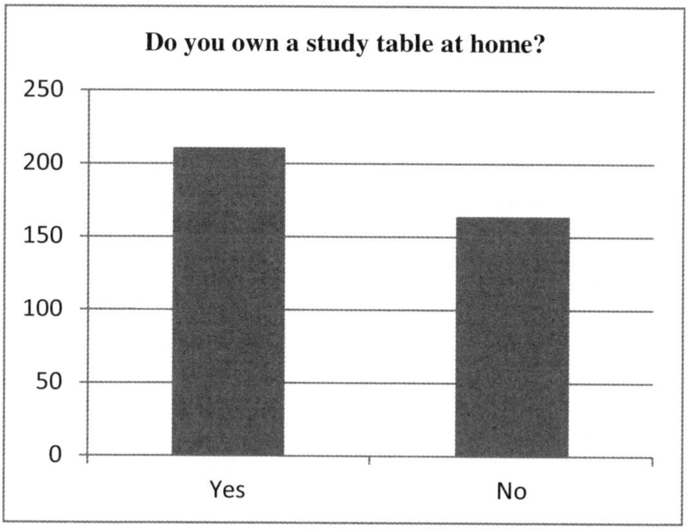

Do you own a study table at home?	n (%)
Yes	211 (56.3)
No	164 (43.7)

Figure 24. Pupil responses to "Do you own a study table at home?"

7 Discussion of Results

7.1 *Classroom behaviours*

Majority of the low attainers were inattentive and easily distracted in class displaying two broad kinds of behaviours: the active disruptive and the passive inattentive kinds of behaviours. Such classroom behaviours were also found in other studies on underachievers in mathematics (Haylock, 1991; Foong, 1999; Van Kraayenoord & Elkins, 2004). Research had links such behaviours to pupils' attitudes and their learning experiences in mathematics. The teachers in this study had also volunteered their reasons as to why the pupils were disruptive and inattentive. Some attributed such behaviours to pupils' bad attitudes and hence leading to their underperformance, others had observed the learning style of the pupils as more towards the kinesthetic type which might explain why such pupils were fidgety and unable to sit and pay attention for long.

Although 63% of the pupils could be good listeners when called upon, majority of them would not seek the help of the teachers when in doubt. Only about 11% of the pupils had good English Language proficiency. Some teachers had attributed such passive behaviours to pupils' shyness or poor command of the English Language.

As for on-task behaviours and the ability to complete classwork in time, the figures showed that only a third of the low attainers were always on-task and gave their best. It could be inferred that due to pupils' easily distracted characteristics majority were not on task nor able to complete their classwork most of the time. Some teachers had made link to pupils' weakness in mathematics, their inability to apply what they had learnt and inability to follow instructions. On the positive note, some teachers had observed that although pupils were slow they could be helped to complete tasks when given more teacher supervision or teacher encouragement to pupils to work harder.

Absenteeism was relatively low for this group of pupils. For those who were always absent (about 8.4%) some teachers had commented that these pupils were always sick and a few could be due to low motivation to want to come to school.

Interestingly, 13 teachers had noted cases of pupils with special needs due to learning disabilities like dyslexia, ADHD or emotive tensions. Some cases were officially diagnosed by school authorities while some were suspected by teachers. This area of special needs or learning disability in mathematics due neurological or psychological issues is significant to be further studied and researched for not many teachers are widely aware or knowledgeable about it to be able to identify such learning problems in their classrooms. Pathology like dyscalculia is still very much an unknown in local mathematics education.

7.2 *Affects in learning mathematics*

Mathematics makes multi-level cognitive demands of the pupils' understanding and application of concepts, from knowing rules and skills to abstract manipulative symbolisms and problem solving. Pupils attempting to memorise mathematics without understanding and rushed into symbolisation or solving convoluted word problems when they are not ready, are likely to form negative attitudes towards the subject especially when they fail to achieve success in assessment. This is evident in the journals of the low attainers in this study. Their feelings of sadness, confusion, hate, boredom, stress and anxiety in learning of mathematics were very much aroused by their experiences that mathematics was hard to understand and they had failed in tests even though they wanted very much to pass exams for the sake of significant others like parents and even grandmother. In the journal writings and drawings of the pupils there were some who expressed happiness about learning and such positive responses could be due to what they thought the teacher would want them to write about mathematics. Among some pupils there was a kind of love-hate feeling, they liked it when they could do well but expressed dislike when they failed to achieve. There was one pupil who wrote "*I love mathematics but I don't know some topic like fraction. I want to be clever but the teacher never explains correctly when I was (in) P3. It is hard for me. That is why I want to learn before I regret.*" Most of the negative responses were frank and reflective. There was a pupil who hated mathematics because "*it was very difficult and needs a lot of thinking*". Such pupils' feelings and feedback are barometers of the kind of teaching and learning that these low attainers were experiencing in the day-to-day mathematics lessons.

From the Pupil Maths Learning Questionnaire, the self-confidence index showed that they were unsure of themselves. 79% of them felt that sometimes they were good at mathematics and 89% reported that they either always or sometimes got stuck when solving mathematics problems. However majority of them reported that they sometimes or never gave up when a problem looked difficult. Probing further through interviews with these low attainers, it found that more than half of them had very low self-concept as they believed they were not good in mathematics. When asked how they could tell that they were not good, half of them said that they had failed their tests and examinations. There was a strong sense of being a failure in mathematics for these young learners. As one pupil said "*No, when you fail, you are no good.*" About 10% of these low self-concept hard core failures thought that they were never good at mathematics and they did not think they could improve. They gave self-depreciating remarks like "*No, I am lazy*" or "*In class, all are good and I am not.*"

Except for this small group, about 90% of the low attainers were positive that they could do better. They believed that if they worked hard, pay more attentive in class and revise more with assessment books at home they could improve. This optimism was attested by their teachers who reported that 77.4% of the low attainers were motivated to learn. Teachers had noted that some pupils tried their best and showed improvement given the appropriate supervision and encouragement.

On the whole the low attainers had positive perception of the value of mathematics. To them mathematics is important and that mathematics was perceived as useful in daily life. 66.5% of the low attainers agreed to the statement "Mathematics is fun.", but only a third agreed that "Mathematics is easy." For obvious reason most of these pupils had experienced failure in the subject. As for perceiving mathematics to be fun it could that at the primary level, the C-P-A (Concrete-Pictorial-Abstract) approach was recommended and these pupils would have been exposed to the use of concrete materials and activities in the learning of early concepts such as whole numbers which they found easy.

7.3 *Home background and support*

In the Pupil Behaviour Questionnaire, some of the teachers volunteered 28 statements relating to pupils' home support when they were asked to give further information to understand pupil attainment. These statements typically were:

- *No family support*
- *No guidance at home / no help from home / has very little home support*

- *On her own at home most of the time. No supervision*
- *Need counselling, many family issues*
- *Have some family support but inconsistent, not free to supervise on a regular basis*
- *Is from a single parent family and consistently shuffles from one babysitter to another*
- *Quite good in Maths. Has support from mother and student care teacher*
- *Excellent home support. Works at a slower pace but gets work done accurately*
- *Good parental support at home*

For low attainers especially if they were mostly from mixed-ability classrooms of class size 30-40 pupils, teachers would appreciate some support at home for these pupils in their homework, as a form of further reinforcement on the learning that had taken place in class. These teachers had noted that good parental support could help some of their low attainers achieve some success even though they were slower in pace. On the other hand, for pupils with behavioural problems in class, they noted that the pupils had little or no family support. Research had also shown that home environment could impact significantly on mathematics achievement and the value of parental support in promoting positive learning outcomes should not be overlooked (Crane, 1999; Leong, 1999; Ng, 2003).

In this study, about a third of the pupils reported that their parents did not check their mathematics work at home and about the same number checked only when they were free. About three quarter of the pupils' main source of help at home was their parents and or siblings. Majority of the parents had either secondary or primary school education. In a local study Leong (1999) found that many parents could not provide direct assistance to their children in the completion of their school work. The parents tried to compensate for this with the employment of tutors for their children. However in this study less than 10% of the pupils reported that they had help from a tutor and about 8% or 30 pupils had no help at home. 21% reported that they sought their siblings for help and another 22% sought both siblings and parents for help. In terms of sibling support, it was noted during the interviews most of these siblings were themselves in primary levels above them, although some were in secondary schools. The quality of home support in helping the low attainers in their mathematics work was questionable.

Majority of the low attainers lived in 4-room or 3-room public housing flats. About 44% of the pupils reported that they did not have a study table of their own. Only about 11% of the pupils used English as the main language at homes with 27% bilingual with English. More than 50% spoke only their mother tongue languages or Mandarin at home. As mathematics is taught in English in Singapore schools, and if the pupils were to have difficulty with the subject, especially in story word problems which are quite pervasive across the topics and levels, the amount and quality of help that family can provide for this group of pupils is doubtful.

8 Conclusion

In conclusion, the results of this study on low attainers' non-cognitive domains are to help in some ways to interpret their underachievement in mathematics. These pupils' affects in learning and their behaviours in the classroom can be seen as barometer of the quality of teaching and learning that these low attainers are experiencing in the day-to-day mathematics lessons. It is not reasonable to infer that pupils bring their attitudes to class that would influence their learning in normal circumstances. For primary school pupils, most likely it is the learning experiences in school that influence their attitudes. Knowing the kind of home backgrounds that the low attaining pupils come from would only make teachers more aware of the educational resources or the lack of which these pupils can or cannot do the work that teachers assigned them for homework. From the expressed feelings of the low attainers about their self-concepts as learners of mathematics and their feelings towards learning the subject, they point to the bad effects of too many tests and assessment in our education system. Pupils see grades from their mathematics tests as information about their ability and therefore judge themselves highly if they did well or lowly if they did not pass. The poor results that the low attainers obtained so frequently are sending the message that they are not good in mathematics and that they are failures. This kind of effect is detrimental to future learning as they set off tension and anxiety in the pupils who already are anticipating difficulties in mathematics.

References

Cheam, F., & Chua, W.L. (2009). Early intervention for pupils at-risk of mathematics difficulties. In K.Y. Wong, P.Y. Lee, B. Kaur, P.Y. Foong, & S.F. Ng. (Eds.), *Mathematics education: The Singapore journey* (pp. 370-389). Singapore: World Scientific.

Crane, J. (1996). Effects of home environment, SES, and maternal test scores on mathematics achievement. *Journal of Educational Research, 89*(5), 305-314.

Denvir, B., Stolz, C., & Brown, M. (1982). *Low attainers in mathematics 5-16: Policies and practices in schools*. London: Methuen Educational.

Foong, P.Y. (1999). Can we identify the reasons for a pupil's low attainment in mathematics? In S.P. Loo (Ed.), *Educational Challenges in the new millennium. Proceedings of the MERA-ERA joint conference* (pp. 648-660). Malacca: Malaysian Educational Research Association.

Haylock, D. (1991). *Teaching mathematics to low attainers, 8-12*. London: Paul Chapman Publishing Ltd.

Leong, K.Y. (1999). *Parental involvement and its effect on academic performance of students.* Unpublished M Ed. Dissertation. National Institute of Education, Singapore.

Leone, C.M., & Richards, H. (1989). Class work and homework in early adolescence: The ecology of achievement. *Journal of Youth and Adolescence, 18*(6), 531-548.

McLeod, D.B. (1992). Research on affect in mathematics education: A reconceptualization. In D. A. Grouws (Ed.), *Handbook of research on mathematics teaching and learning* (pp. 575-596). New York: Macmillan.

Ministry of Education. (2006). *Mathematics syllabus — Primary*. Singapore: Author.

Mullis, I.V.S., Martin, M., & Foy, P. (2008). TIMSS 2007 International mathematics report: Findings from IEA's Trends in International Mathematics and Science Study at the fourth and eighth grades. TIMSS & PIRLS International Study Center, Boston College, United States.

Ng, T.J. (2003). *Possible factors relating to poor performance of underachievers in primary schools.* Unpublished M Ed. Dissertation. National Institute of Education, Singapore.

Schoenfeld, A.H. (1989). Explorations of students' mathematical beliefs and behavior *Journal for Research in Mathematics Education, 20*, 338-355.

Van Kraayenoord, C. E., & Elkins, J. (2004). Learning difficulties in numeracy in Australia. *Journal of Learning Disabilities, 37*(1), 32-41.

Wong, K.Y., & Quek, K.S. (2009). Enhancing mathematics performance (EMP) of mathematically weak pupils. *Final Research Report for Project No. CRP 47/03 WKY*. Singapore: NIE CRPP-NTU.

Chapter 4

Learning Experiences of Low Attainers

Berinderjeet KAUR Masura GHANI

This chapter explores the learning experiences of 364 Primary 4 low attainers in mathematics. Data was collected from pupils using a questionnaire and an interview about their learning experiences related to mathematics lessons in school. An innovative method, using pictures as stimulus, was adopted to engage pupils to talk about their lessons. From the questionnaire and interview data it is apparent that there was a mismatch between how pupils were taught and preferred to be taught. Almost all the pupils experienced teacher-led whole class instruction during their mathematics lessons. A study of ten teacher-led whole class instruction lessons showed that these lessons were not unique but had some commonalities. The mathematical tasks used during instruction were routine and repetitive. Teachers also did not stimulate the development of pupils' metacognition.

1 Introduction

Failure in mathematics due to factors that may be controlled is unjust particularly for low attaining in mathematics pupils. One such factor is the learning experience that results from classroom instruction provided by teachers. Knowledge of preferred learning experiences by these pupils is therefore pertinent if teachers are to provide them with experiences that would enhance their learning and also develop positive attitudes towards mathematics. Therefore, in this chapter we draw on the data of the study and explore from the perspective of the pupils how they were taught and would like to be taught. We also examine the lessons of their teachers. This chapter answers the research question: "What are the learning experiences of low attaining in mathematics pupils in school?"

2 Review of Literature

The review of literature in this chapter is specific to the learning experiences of low attaining in mathematics pupils. According to Reusser (2000) there is sufficient evidence in research on mathematics learning and teaching that most observed failures and substandard performances are due to deficiencies in the teaching and learning environments rather than genetic factors. Reusser in his review of theoretical and empirical research, from a cognitive instructional perspective, contends that an effective teaching environment positively impacts students' mathematics attainment levels regardless of grade levels or mathematical ability. His perspective of an "effective teaching environment" for low attainers centres around adaptivity and empathy in teaching. He recommends the use of micro adaptation — moment-to-moment decisions of teachers aiming at tailoring instruction to the needs of different learners.

Direct structured instruction has also been found to be effective with students having difficulties in mathematics (Harris, Miller, & Mercer, 1995; Jitendra & Hoff, 1996; Van Luit, 1994; Wilson, Majsterek, & Simmons, 1996). Direct instruction is systematic explicit instruction which is teacher-led (Jones, Wilson, & Bhojwani, 1997) and generally follows a fixed pattern of actions (Archer & Isaacson, 1989). Kroesbergen and Van Luit (2002) detail a typical direct instruction lesson as having three phases. In the opening phase the students' attention is gained, previous lessons are reviewed and the goals of the lesson are stated. In the main part of the lesson the teacher demonstrates how a particular task can be solved following which the students and teacher work together on a few more similar tasks. When the students appear to have sufficient understanding of the tasks they are given new tasks to practice independently. The teacher monitors the students during such practice and provides feedback on completed tasks.

Cardelle-Elawar (1995) studied the effects of metacognitive instruction on low achieving students' mathematics achievement and their attitude toward mathematics. The subjects of his study were third to eighth grade students. The results of the study indicated that metacognitive training for problem solving was effective in facilitating the problem solving processes of low achieving students. Students who went through the intervention became more critical of themselves. They also showed metacognitive potential when stimulated by explicit individualised instruction. The main recommendations of this study are i) special consideration should be given to individual student's uniqueness, strengths and weaknesses; ii) these students need a supportive atmosphere in which errors and mistakes are considered a source of learning and not an occasion for punishment; iii) these students need more structure in the classroom; and iv) these students warrant a great deal of interaction between teacher and student.

One of the contentious topics of study is whether low attainers can / should be taught higher order thinking skills which are normally associated with high achievers. Watson (2001), in her study of low attaining Year 9 students found that students were able to make shifts in their thinking from the superficial features of mathematical tasks to forms of mathematical thought. She cites a specific example where students were able to shift from seeing fractions as congruent shapes to seeing fractions as quantities using the idea of areas to make the link. Watson asserts that low attainment is not the result of an inability to think but the lack of structured work that promotes higher order thinking among low attainers. Zohar and Dori (2003) in their study investigated whether low achieving students can gain from teaching and learning processes that are designed to foster higher order thinking skills. The findings of the study showed that both high achievers as well as the low achievers, made considerable progress with respect to their initial scores. They suggest that teachers should encourage students of all levels to engage in tasks that involve higher order thinking skills.

3 Method

In this chapter we present the method which was part of the Low Attianers in Primary Mathematics (LAPM) study that is specific to the research question explored in this chapter.

3.1 *Subjects*

Although 375 Primary 4 pupils participated in the LAPM study, not all were present on the days the interviews were conducted in schools. Therefore only 346 were interviewed. These pupils also completed the Pupil Maths Learning Questionnaire prior to the interview session. Some pupils who were absent for the interviews, did manage to complete the questionnaire during our visits to the school, for the teacher lesson observation. Therefore in total 364 pupils completed the questionnaire. Also as part of the study, the lessons of ten teachers who taught the pupils were observed by the research team.

3.2 *Instruments*

Data on learning experiences of the pupils were collected from the pupils using a questionnaire and an interview. The lessons of teachers in the study were also observed for learning experiences data.

3.2.1 Pupil maths learning questionnaire

The Pupil Maths Learning Questionnaire, shown in Appendix G, comprised items concerning home support, teachers' classroom pedagogy, perceptions about mathematics, self-confidence and learning preferences. In this chapter we focus on the items which are related to teachers' classroom pedagogy and pupils' learning preference. There were four statements on teachers' classroom pedagogy (P) and ten statements on pupils' learning preference (LP). Specific to the research question addressed in this chapter, we will only explore pupil responses to the following items:

P1: My maths teacher always teaches maths in the same way. She stands in front and teaches.
P2: We do a lot of group activities in our maths lessons.
LP1: I like to learn maths by doing activities.
LP2: I like to know how maths is used in my daily life.
LP3: I like to work in groups during maths lessons.
LP4: Diagrams and figures help me learn maths better.
LP5: The computer helps me learn maths better.

3.2.2 Pupil interviews

From our interactions with the pupils and their teachers in the project we found that pupils lacked the language to talk about their learning experiences and actions of their teachers during mathematics lessons. So, to facilitate pupils' talk about their learning experiences we adopted an idea from child psychiatry about using pictures as stimulus for interviews (Angold, 1976). We also found the pupils rather reserved in their oral communication with us when they were in a one-to-one interview setting. This observation led us to adopt a group interview format for our study. Pupils were interviewed in groups of 4 to 5 persons. During the interviews the researchers of the study used four pictures of mathematics lessons to stimulate talk amongst the pupils about how their math teachers usually taught them in class and what their preferences for learning mathematics were. The four pictures shown in Figure 1 were used for the interviews. The researchers made it known to the pupils that the teacher is present in all four pictures to facilitate their learning.

The prompts used for the interviews, shown in Appendix H, belonged to three categories, mathematics lessons, homework and self. In this chapter we only focus on the prompts related to mathematics lessons. The four pictures A, B, C and D were put on the table around which the pupils and interviewer sat for the interview session. The following prompts were used to engage pupils in talking about their mathematics lessons at school.

Prompt 1: Which picture shows the way your maths teacher usually teaches you in class?

Prompt 2: Which class do you want to be in? Why?
Prompt 3: Which class don't you want to be in? Why?

Picture A	Picture B
	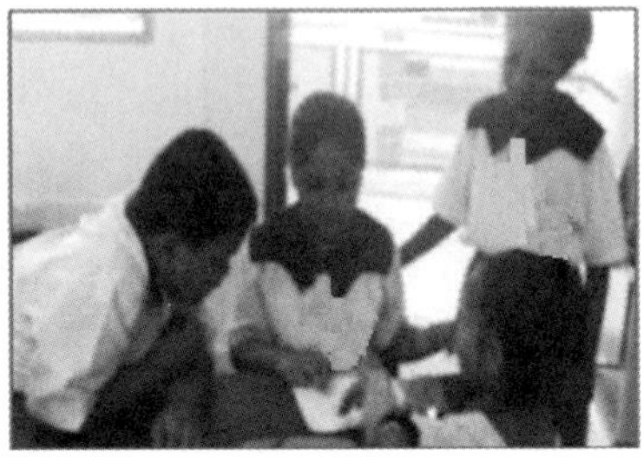
Teacher-led whole class instruction	Group work (pupils working on tasks without manipulatives)
Picture C	Picture D

Individual working on task with manipulatives	Group work (pupils working on tasks with manipulatives)

Figure 1. Pictures of mathematics lessons

3.2.3 Lesson observations

Nine schools participated in the project. The lesson of one teacher per school who welcomed the researchers to his / her class was observed. In one school, two teachers volunteered and therefore a total of ten lessons were observed. The teachers taught mathematics to pupils participating in the project. Our lesson observations were guided by 10 analytical questions shown in Appendix J.

4 Data and Findings

In this section we first present the data and findings of the pupil questionnaire followed by the interviews according to the three interview prompts chronologically. Next we present our analysis of the ten lessons that were observed as part of the study.

4.1 *Pupil maths learning questionnaire*

The data collected via the Pupil Maths Learning Questionnaire is shown in Table 1. Majority of the pupils (82.7%) indicated that their teachers used teacher-led whole class instruction to teach them during mathematics lessons. About half of the pupils (54.1%) reported that they did group activities during mathematics lessons. A high percentage of the pupils preferred to do activities (88.2%) and group work (80.5%) during mathematics lessons. 77.2 % of the pupils also indicated that the use of diagrams and figures (representations) helped them learn mathematics better. A large proportion of the pupils (70.3%) also reported that the computer helped them learn mathematics better. Many pupils (79.4%) were also curious to know how mathematics would be applicable in their daily lives.

Table 1
Responses to the pupil maths learning questionnaire items

No.	Statements	Frequency n (%)	
		Yes	No
P1	My maths teacher always teaches maths in the same way. She stands in front and teaches.	301 (82.7)	63 (17.3)
P2	We do a lot of group activities in our maths lessons.	197 (54.1)	167 (45.9)
LP1	I like to learn maths by doing activities.	321 (88.2)	43 (11.8)
LP2	I like to know how math is used in my daily life.	289 (79.4)	75 (20.6)
LP3	I like to work in groups during maths lessons.	293 (80.5)	71 (19.5)
LP4	Diagrams and figures help me learn maths better.	281 (77.2)	83 (22.8)
LP5	The computer helps me learn maths better.	256 (70.3)	108 (29.7)

4.2 *Pupil interview*

4.2.1 Prompt 1: Which picture shows the way your Maths teacher usually teaches in class?

98% of the pupils interviewed said that their teachers always used whole class instruction for mathematics lessons. 50% of the pupils interviewed also reported that sometimes they did group work during mathematics lessons.

4.2.2 Prompt 2: Which class do you want to be in? Why?

Figure 2 shows the preference of the pupils with regards to the type of instruction, as shown in the pictures, they desired. The highest percentage, i.e., 40%, of the pupils said that they preferred to work in groups on mathematical tasks with manipulatives during mathematics lessons while 22% of the pupils preferred to work in groups on tasks without manipulatives. It must be noted that during the interviews the researchers explained explicitly the differences between pictures B and D. Less than one third of the pupils preferred to be in a lesson where whole class instruction was the norm. Interestingly, only 10% of the pupils indicated preference for individual work with manipulatives. Figures 3, 4, 5 and 6, show representative samples of all the reasons given by the pupils in support for their preferences.

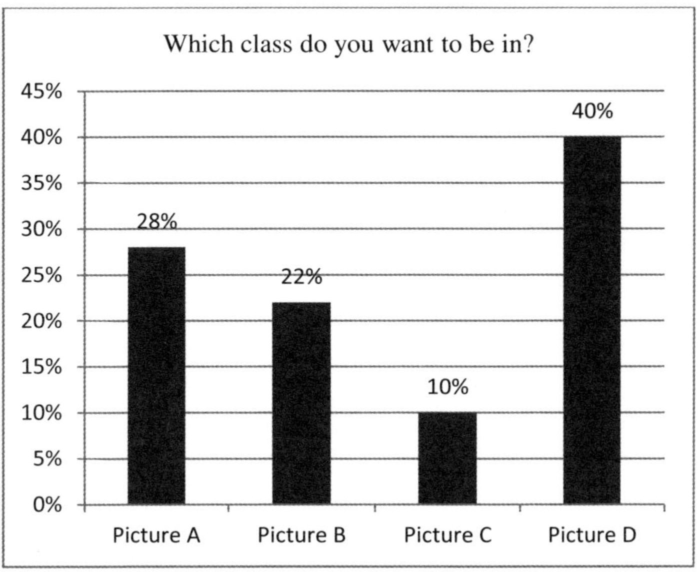

Figure 2. Responses to *"Which class do you want to be in?"*

- *I like to study in a quiet classroom rather than do activity.*
- *We need to listen to teacher. We will know what to do for exam.*
- *Teacher can teach us more than we can learn ourselves.*
- *I like to be in a quiet class when learning.*
- *I can be better as the teacher is teaching me everything.*
- *Teacher can teach, we can listen.*
- *I can understand better when the teacher teach*
- *Teacher can teach and we can listen and we will be able to pass the exam.*
- *It is easier to understand when teacher is teaching on the board.*
- *I can learn better.*

Figure 3. Sample of pupils' responses to why they want to be in class shown by Picture A (Teacher-led whole class instruction)

From the sample responses in Figure 3, it is apparent that pupils preferred whole class instruction as the class was i) quiet and conducive for learning; and ii) the teacher explained and demonstrated how to do the mathematics and they perceived this will help them pass exams.

- *I enjoy discussing with my friends.*
- *I can finish my work quickly when discussing with my friends.*
- *I can check work with my friend if I get it wrong.*
- *I can discuss and learn more.*
- *I can get the answer easily when work in groups.*
- *I like to discuss with my friends.*
- *My friends can help me when things not clear.*
- *Many people and they can help me.*
- *I like to discuss Mathematics with friends. Help me understand and enjoy learning.*
- *When we discuss with friends, it is easier.*

Figure 4. Sample of pupils' responses to why they want to be in class shown by Picture B (Group work — pupils working on tasks without manipulatives)

From the sample responses in Figure 4, it is apparent that pupils preferred to work on tasks in groups without manipulatives as they found peer support helpful to i) discuss their work, ii) clarify their doubts, iii) seek help when necessary, iv) check their work for correctness, and v) finish their work promptly.

- *I can learn better when learning alone.*
- *I like to be alone. When in group, very confusing.*
- *If I get something wrong, I can learn from my own mistake.*
- *We must try to do individually as in exam we need to do individually.*
- *Very interesting to do activity individually.*
- *I can learn maths if I do on my own. I like to work alone.*
- *I can do alone and later teacher will be proud of me.*
- *When we are alone, we can do more Maths.*
- *I like to solve problem by myself.*
- *I like to learn on my own. I don't like people telling me how to do things.*

Figure 5. Sample of pupils' responses to why they want to be in class shown by Picture C (Individual working on tasks with manipulatives)

From the sample responses in Figure 5 it is apparent that pupils preferred to work on tasks with manipulatives by themselves as they can learn at their own pace in a quiet environment similar to that of an examination.

- *When we do the activity, we can feel that maths is fun.*
- *Able to discuss and work together with materials.*
- *Can make more friends and learn better.*
- *More fun to work with a group, can discuss with friends if I am not sure.*
- *I learn better when I am allowed to do things in class.*
- *Group activities are fun.*
- *I feel happy when I can do things to help me understand and improve.*
- *Fun to do activities.*
- *With activities, mathematics is fun.*
- *We can see how things happen and touch things.*

Figure 6. Sample of pupils' responses to why they want to be in class shown by Picture D (Group work — pupils working on tasks with manipulatives)

From the sample responses in Figure 6 it is apparent that pupils preferred to work on tasks in groups with manipulatives as they found i) interacting with peers a fun and good way to learn, and ii) the "hands on" experience gratifying and a meaningful way of learning.

4.2.3 Prompt 3: Which class don't you want to be in? Why?

Figure 7 shows the preference of the pupils with regards to the type of instruction, as shown in the pictures, they disliked. More than half of the pupils (62%) preferred not to be in a class where they would have to work by themselves on a task with manipulatives (Picture C). This is consistent with the low percentage of pupils who chose Picture C for Prompt 2. Also, less than 10% of the pupils preferred not to be in class where there would work in groups on a task with manipulatives. This is again consistent with the high percentage of pupils who chose Picture D for Prompt 2. Figures 8, 9, 10 and 11, show representative samples of all the reasons given by the pupils in support of not wanting to be in a particular classroom setting.

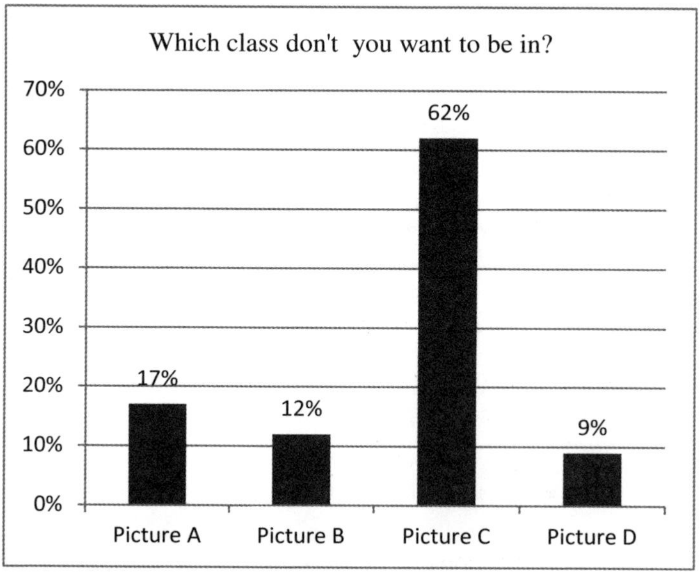

Figure 7. Responses to *"Which class don't you want to be in?"*

- *It is boring and makes me sleepy.*
- *Very confusing and cannot understand.*
- *Looks boring and hate copying from the board.*
- *Sometimes don't understand when teacher teach.*
- *Boring as I need to pay attention all the time.*
- *Teacher asks many questions and I can't do things my way.*
- *When teacher is talking, we feel sleepy.*
- *Boring when teacher is standing in front at the board.*
- *Boring got nothing to do but copy.*
- *No group work if teacher is teaching all the time.*

Figure 8. Sample of pupils' responses to why they don't want to be in class shown by Picture A (Teacher-led whole class instruction)

From the sample responses in Figure 8, it is apparent that pupils found teacher-led whole class instruction as i) boring and difficult to understand; and ii) rigid due to the lack of opportunities to discuss with peers.

- *Others may make mistake and may be noisy.*
- *Too many people may quarrel.*
- *Discussion only will not learn anything for exam.*
- *I don't like to be in a noisy environment.*
- *Everyone will be talking, playing and not doing the activity.*
- *We might fight when discussing and write the answer anyhow.*
- *Sometimes they talk loudly and not about maths.*
- *Other people give answer, later exam I don't know how to do.*
- *Cannot concentrate*
- *They don't let me do, keep pulling the paper.*

Figure 9. Sample of pupils' responses to why they don't want to be in class shown by Picture B (Group work — pupils working on tasks without manipulatives)

From the sample responses in Figure 9, it is apparent that pupils found group discussions without manipulatives to be distracting which may lead to several unpleasant situations.

- *Work alone may not know how to do and then do the wrong thing.*
- *It is boring and lonely when doing by ourselves.*
- *Lonely to carry out activities alone.*
- *If alone cannot study well, cannot ask anybody about the activity.*
- *I don't want to do activity alone.*
- *No one to help if I don't know how to do.*
- *If I don't understand, I will do it wrongly.*
- *Scared if I don't understand what teacher wants.*
- *Don't want to do things alone. With other people we can do better.*
- *No one to help and need to solve ourselves.*

Figure 10. Sample of pupils' responses to why they don't want to be in class shown by Picture C (Individual working on tasks with manipulatives)

From the sample responses in Figure 10, it is apparent that pupils preferred not to work by themselves on tasks with manipulatives as they i) lacked the confidence to attempt tasks without the support of peers and teachers; and ii) felt bored and lonely working by on their own.

- *Work cannot be done properly in a group.*
- *Don't like to work in groups. Later I will make mistakes.*
- *If we make too much noise, teacher will ask us to stop the activity.*
- *When they make noise cannot hear what teacher is saying.*
- *If messy, teacher will scold.*
- *Holding things are difficulty. If I do wrongly, my friends scold me.*
- *Will make a mess.*
- *More difficult as the activity may be messy.*
- *If I don't do properly, teacher will scold.*
- *Very noisy and messy.*

Figure 11. Sample of pupils' responses to why they don't want to be in class shown by Picture D (Group work — pupils working on tasks with manipulatives)

From the sample responses in Figure 11, it is apparent that pupils found working in groups with manipulatives usually resulting in a noisy learning environment, with them making errors and being scolded by their teachers and peers.

4.3 *Lessons observed*

Each lesson was observed by at least two researchers. Following the observation, a reflection of the lesson was guided by the analytical questions that provided the theoretical lens for analysis. The questions were as follows:

1a. What was the focus of the lesson — creation or dissemination of knowledge?
1b. What was the instructional sequence of the lesson like?
2. Did the teacher tailor instruction to meet the needs of different learners?
3. What were the characteristics of mathematical tasks used in the lesson?
4a. Did the teacher ask pupils questions (both open and closed)?
4b. Did the teacher praise pupils for their responses?
4c. Did the teacher build on the pupils' responses (both correct and incorrect) to explain mathematical concepts?
4d. Did the teacher encourage pupils to ask questions?
4e. Did the pupils ask the teacher any questions to clarify their doubts?
4f. Did the pupils ask their peers any questions to clarify ideas?

The main aspects of the lessons that the researchers concurred on are presented in Tables 3, 4, 5, 6 and 7. An analysis of the data in Tables 3, 4, 5, 6 and 7 was carried out according to the ten analytical questions.

Table 3

Data for lessons of Teachers A and B

AQ	Teacher A (School 1) Topic: Time [duration] N = 35	Teacher B (School 2) Topic: Symmetry N = 38
1a	Focus: dissemination of knowledge	Focus: dissemination of knowledge
1b	i) Introductory phase — mention that lesson was on duration ii) Main phase — application of knowledge (varied examples worked on the board with inputs from pupils) iii) Consolidation phase — competition between boys and girls on conversion from 12hour to 24hour clock and vice versa followed by pair work on new tasks. Teacher provided between desk instructions.	i) Introductory phase — mention that lesson was on a new topic symmetry. ii) Main phase — demonstration of concept using video clips followed by individual "hands on" work by the pupils — identifying line of symmetry of different shapes. iii) Consolidation phase — pupils worked individually on new tasks, while teacher and allied educator provided between desk instructions.
2	No apparent attempt	No apparent attempt
3	Routine	Routine and repetitive
4a	Teacher asked closed questions such as "What do we do here?", "What's the next step?"	Teacher asked closed questions such as "Which one is the answer?", "What is the next step?".
4b	Teacher praised individual pupil for their responses to her questions.	Teacher praised pupils when correct choral responses were given.
4c	Teacher drew input from pupils and used only the correct responses from pupils to solve the examples during board work.	Teacher drew input from pupils and used only the correct responses to develop the concept of symmetry.
4d	Teacher used a general phrase "Any questions?" to encourage pupils to ask questions.	Teacher used a general phrase "Any questions?" to encourage pupils to ask questions.
4e	Pupils raised their hands during seatwork to ask the teacher questions.	Pupils raised their hands during seatwork to ask the teacher and allied educator questions.
4f	Teacher created opportunity for peers to clarify ideas through Rally Coach activity. However, the questions were too difficult for the pupils to do so.	Pupils clarified their ideas with the peers during seatwork.

*N represents the number of pupils in the class.

Table 4

Data for lessons of Teachers C and D

AQ	Teacher C (School 3) Topic: Time [duration] N = 11	Teacher D (School 4) Topic: Time [duration] N = 35
1a	Focus: creation of knowledge	Focus: dissemination of knowledge
1b	i) Introductory phase — review of past lesson and use of real life contexts to arouse pupils' interest. ii) Main phase — development of concept and application of knowledge (adequate examples worked on the board with inputs from pupils) iii) Consolidation phase — pupils worked individually on new tasks, while teacher provided between desk instructions and feedback on completed tasks.	i) Introductory phase — review of past lesson and use of real life contexts to arouse pupils' interest. ii) Main phase — development of concept and application of knowledge with the help of a physical clock. iii) Consolidation phase — pupils worked individually on new tasks, while teacher and allied educator provided between desk instructions.
2	No apparent attempt	No apparent attempt
3	Routine and repetitive	Routine
4a	Teacher asked both closed and open questions such as "What time are we supposed to start the lesson?", "How do you know we have only 50minutes left?"	Teacher asked closed questions such as "What time do we have to be in the quadrangle?", "Is 7.15am in 12hour or 24hour clock?"
4b	Teacher praised the pupils who completed the given work and gave them a new worksheet to work on.	Teacher praised individual pupil during seatwork.
4c	Teacher drew input from pupils and used the responses from pupils to solve the examples during board work.	Teacher used pupils' responses to show various ways of dividing the timeline in solving the first example.
4d	Teacher used a general phrase "Any questions?" to encourage pupils to ask questions.	Teacher used a general phrase "Any questions?" to encourage pupils to ask questions.
4e	Pupils raised their hands during seatwork to ask teacher.	Pupils raised their hands during seatwork to ask the teacher and allied educator questions.
4f	Pupils had no opportunity to clarify ideas with their peers as individual seatwork was given.	Pupils had no opportunity to clarify ideas with their peers as individual seat work was given.

*N represents the number of pupils in the class.

Table 5

Data for lessons of Teachers E and F

AQ	Teacher E (School 5) Topic: Symmetry N = 22	Teacher F (School 5) Topic: Tessellation N = 19
1a	Focus: dissemination of knowledge	Focus: dissemination of knowledge
1b	i) Introductory phase — mentioned that lesson was on a new topic — symmetry. ii) Main phase — demonstration of concept of symmetry using software (intellimath) followed application of knowledge through board work with inputs from pupils. iii) Consolidation phase — pupils worked individually on similarly tasks	i) Introductory phase — mentioned that lesson was on a new topic — tessellations. ii) Main phase — demonstration of concept of tessellation via examples and non-examples. Pupils worked in groups with unit shapes to make tessellated patterns and teacher encouraged peer evaluation. iii) Consolidation phase — pupils worked in pairs and again were given unit shape to make tessellated patterns.
2	No apparent attempt	No apparent attempt
3	Routine and repetitive	Routine and repetitive
4a	Teacher asked both closed and open questions such as "Is it symmetric?", "What is the difference in the lines of *A* and *D*?"	Teacher asked both closed and open questions such as "Where is the unit shape?", "Can the pattern can be tessellated?", "What do you think?"
4b	Teacher praised individual pupil during seatwork.	Teacher praised pupils for their comments on their peers' answers.
4c	Teacher used pupils' correct choral response to explain the concept of symmetric figure. When there were different responses, teacher drew input from pupils with the correct answer to show the solution to the whole class.	Teacher repeated, paraphrased the pupils' answers and used them to further explain the concept of tessellation. When one of the group presented a wrong answer, the teacher asked the class for an explanation on why they think the answer is wrong and used the correct reasons to further emphasis the concept of tessellation.
4d	Teacher used a general phrase "Any questions?" to encourage pupils to ask questions.	Teacher used a general phrase "Any questions?" to encourage pupils to ask questions.
4e	Pupils raised their hands during individual seatwork to ask the teacher questions.	Pupils raised their hands during group activity to ask the teacher questions.
4f	Pupils clarified their doubts with their peers during seatwork.	Pupils clarified their doubts with their peers during group activity.

*N represents the number of pupils in the class.

Table 6

Data for lessons of Teachers G and H

AQ	Teacher G (School 6) Topic: Time [word problems] N = 18	Teacher H (School 7) Topic: Time [discussion of homework] N = 6
1a	Focus: dissemination of knowledge	Focus: dissemination of knowledge
1b	i) Introductory phase — review previous lesson and stated the goal of the lesson ii) Main phase — application of previous knowledge through board work with inputs from the pupils iii) Consolidation phase — pupils were allowed to work individually or in pairs on new tasks.	i) Introductory phase — mention that the lesson will be on discussion of homework questions. ii) Main phase — demonstration of concept with the help of a physical clock followed by discussion of homework questions. iii) Consolidation phase — not evident
2	No apparent attempt	No apparent attempt
3	Routine	Routine
4a	Teacher asked closed questions such as "Is this in 12hour or 24hour clock?", "Do I add or subtract?", "What do I do next?"	Teacher asked closed questions such as "How many hours in a day?", "How many minutes in 2h 15mins?"
4b	Teacher praised individual pupil during seatwork.	Teacher praised the pupils for doing individual "hands-on" work correctly.
4c	Teacher drew input from pupils and used only the correct responses from pupils to solve the examples during board work.	During between desk instruction, teacher used one of the pupil's answer to do a revision with the class on another method to convert minutes to hours. Teacher also used individual pupil's board work to explain the steps involved in solving the questions.
4d	Teacher used a general phrase "Any questions?" to encourage pupils to ask questions.	Teacher didn't encourage pupils to ask questions as she was able to check that every pupil understood the lesson.
4e	Pupils raised their hands during seatwork to ask the teacher questions.	Teacher walked around and checked every single pupil's work.
4f	Pupils clarified their doubts with their peers during seatwork.	Pupils didn't need to clarify their doubts with their peers as teacher was able to help them individually.

*N represents the number of pupils in the class.

Table 7

Data for lessons of Teachers I and J

AQ	Teacher I (School 8) Topic: Symmetry N = 37	Teacher J (School 9) Topic: Tessellations N = 38
1a	Focus: dissemination of knowledge	Focus: dissemination of knowledge
1b	i) Introductory phase — mentioned past lesson completed the topic time and stated the goal of the present topic. ii) Main phase — demonstration of concept using video-clips and cut-outs of the alphabet by the teacher followed by "hands on" work by pupils in groups — identifying the lines of symmetry of the alphabet. iii) Consolidation phase- pupils worked individually on similar tasks.	i) Introductory phase — mentioned that the goal of the present topic is to explore something new on tessellations and reviewed past lesson on tessellations. ii) Main phase — demonstration of common difficulty that pupils faced when doing tessellations followed by "hands on" work by pupils in groups — forming tessellated patterns. iii) Consolidation phase — pupils worked individually on similar tasks
2	No apparent attempt	No apparent attempt
3	Routine and repetitive	Routine and repetitive
4a	Teacher asked closed questions such as "Where is the line of symmetry?", "Is this figure symmetrical?"	Teacher asked closed questions such as "What are the 3 things about tessellation?", "Is this a tessellation?"
4b	Teacher praised the pupils for completing the group work on time.	Teacher praised the pupils for presenting their group work.
4c	During whole class instruction, one of the pupils said that the line of symmetry is always from top (vertical), the teacher explained that's not always the case and showed the class the next example where the line of symmetry is horizontal. Teacher also used pupils' response to explain the uniqueness in the number the lines of symmetry in the letters "O" and "K".	Teacher used pupil's presentation to explain the concept of using the same unit shape to form different tessellations. Teacher also used one pupil's work to share with the class a good way to tessellate with a fixed amount of unit shape — numbering each unit shape.
4d	Teacher used a general phrase "Any questions?" to encourage pupils to ask questions.	Teacher asked pupils to raise their hands if they have any questions.
4e	Pupils raised their hands during group activity to ask the teacher questions.	Pupils raised their hands during group activity and individual seatwork to ask the teacher questions.
4f	Pupils clarified their doubts with their peers during group activity.	Pupils had no opportunity to clarify ideas with their peers as individual seat work was given.

*N represents the number of pupils in the class.

4.3.1 What were the foci of the lessons?

All the ten lessons observed depicted teacher-led whole class instruction. In nine of the lessons, the teachers mainly disseminated knowledge while in one the teacher engaged pupils in creation of knowledge.

4.3.2 What were the instructional sequences of the lessons like?

Nine of the ten lessons observed had three phases in the instructional sequence: the introductory, main and consolidation phases. The consolidation phase was not evident in one of the ten lessons. During the introductory phase, only three teachers reviewed what the pupils had learnt in their previous lessons. Two out of the three teachers also used real life context to arouse their pupils' interest. The other seven teachers only mentioned what the pupils would be learning during the lessons.

In the main phase of the lessons, only two of the ten teachers focused on the development of a mathematical concept before moving onto the application of knowledge using adequate examples worked on the board. Two of the teachers also used a manipulative (a physical clock) to show the pupils how to apply the knowledge. Six of the teachers demonstrated a mathematical concept using teaching aids such as video clips and cut-outs of alphabets before proceeding to the application of the knowledge through "hands on" work by the pupils. The remaining two teachers just focused on the application of knowledge using varied examples they worked on the board with inputs from the pupils.

During the consolidation phase in six of the ten lessons observed, pupils working individually on tasks set by their teachers. In three of these lessons the tasks were new tasks while in the other three the tasks were similar to the examples discussed in the main phase of the lessons. In two other lessons pupils worked in pairs on new tasks while in one other lesson pupils worked in pairs on a task similar to the one their teacher discussed in the main phase. In one of the lessons, the teacher had a competition as part of the consolidation phase before transiting the pupils to work in pairs on a new task.

4.3.3 Did the teachers tailor instruction to meet the needs of different learners?

During whole class instruction, there were no apparent attempts made by the teachers to tailor instruction to meet the needs of learners with differing ability. All ten lessons observed catered to pupils of average ability. In four of the lessons observed, the teachers helped the low ability pupils during seatwork.

4.3.4 What were the characteristics of mathematical tasks used by the teachers in their lessons?

The tasks used by the teachers were routine and repetitive and it was apparent that the teachers made no attempt to engage pupils in higher order thinking.

4.3.5 Did the teachers ask pupils questions (both open and closed)?

Majority of the teacher, i.e. seven, asked their pupils closed questions only such as "What is the next step?" and "What do I do next?" Only three teachers asked both closed and open questions such as "What time are we suppose to the start lesson?" and "How do you know we have fifty minutes left?"

4.3.6 Did the teachers praise pupils for their responses?

All the teachers praised their pupils either individually or in groups for completing given work and answering questions. In one lesson, the teacher engaged the pupils in peer evaluation and commended them for their attempt to give comments to their peers.

4.3.7 Did the teachers build on the pupils' responses (both correct and incorrect) to explain mathematical concepts?

All the teachers drew inputs from their pupils while working on the boards demonstrating the application of knowledge to solve mathematical tasks. However, only seven of the ten teachers used only the correct responses from the pupils to explain mathematical concepts. The other three teachers used individual responses to clarify misconceptions as well as reinforce the concept being taught. For example, one of the pupils in Teacher I's classroom made a remark that a line of symmetry is always *"from the top"*. Teacher I corrected the misconception by showing an example where the line of symmetry was horizontal. She also used her pupils' answers to show the uniqueness in the number of lines of symmetry in the letters "O" and "K".

4.3.8 Did the teachers encourage pupils to ask questions?

Eight of the ten teachers used a general phrase "Any questions?" to encourage their pupils to ask questions. In one class, where there were only 6 pupils the teacher did not encourage her pupils to ask questions but she went from pupil to pupil attending to their difficulties. In another classroom, the teacher asked her pupils to raise their hands if they had questions.

4.3.9 Did the pupils ask their teachers any questions to clarify their doubts?

None of the pupils in the ten lessons observed asked their teachers any question to clarify their doubts during whole class instruction. Instead the pupils clarified their doubts with their teachers by raising their hands during groupwork and individual seatwork.

4.3.10 Did the pupils ask their peers any questions to clarify their ideas?

In only four of the ten lessons observed did pupils have opportunities to clarify their ideas with their peers.

5 Discussion and Concluding Remarks

From the Pupil Maths Learning Questionnaire data, it is apparent that majority of the pupils (82.7%) in the study experienced teacher-led whole class instruction during mathematics lessons. 54.1% of the pupils also reported that they did a lot of group activities during mathematics lesson. 88.2% and 80.5% of the pupils preferred to do activities and group work respectively during mathematics lessons. More than 70% of the pupils also indicated that the use of aids such as diagrams, figures and computers helped them in learning mathematics. 79.4% of the pupils were curious to know how the mathematics they learnt would be used in their daily lives. The questionnaire data highlights a mismatch in the teachers' classroom pedagogy and the pupils learning preferences.

The interview data of the pupils in the study reveals that 98% of them are taught mathematics in classrooms where teacher-led whole class instruction is the norm. At times in some of the classrooms teachers do carry out group work during which pupils work on tasks without any manipulatives. However, teacher-led whole class instruction was the preference of only 28% of the pupils in the study. This finding is coherent with the data from the pupils' survey. 40% of the pupils preferred to work in groups on mathematical tasks with the help of manipulatives, while another 22% of the pupils preferred to work in groups on mathematical tasks in the absence of manipulatives. Pupils preferred to work on tasks in groups with manipulatives as they found interacting with peers a fun and good way to learn, and the "hands on" experience gratifying and meaningful in learning. From the above findings the apparent mismatch between how teachers teach these pupils and how these pupils would like to be taught in mathematics lessons may explain partially the low attainment in mathematics of these pupils. This finding reinforces that of Reusser (2000) that most observed failures and substandard performances are due to deficiencies in the teaching and learning environments rather than genetic factors.

The ten lessons observed depicted teacher-led whole class instruction. In one lesson the teacher engaged pupils in the creation of knowledge while in the other nine lessons teachers merely disseminated knowledge to the pupils. The first two phases of all the ten lessons were similar to that of Kroesbergen and Van Luit (2002), i.e., pupils' attention was gained and teacher demonstrated how a particular task can be solved. However, in the third phase teachers lacked in ensuring that pupils had sufficient understanding of the tasks before they were given more tasks to practice independently. In one of the lessons the third phase was not evident. All ten lessons observed catered to pupils of average ability. Four of the ten teachers, whose lessons were observed, used between desk instructions to help the low ability pupils during seatwork. The tasks used by the teachers were routine and repetitive and it appears that teachers made no attempt to engage pupils in higher order thinking. In addition teachers asked mainly closed questions. Watson (2001) and Zohar and Dori (2003) have found that low attaining pupils are capable of making shifts in their thinking and improving in their mathematics attainment when challenged with higher order thinking tasks.

In the ten lessons observed, generally the learning environments were conducive. Teachers were welcoming of mistakes, praising pupils for good effort, encouraging pupils to ask questions, but rarely engaging them in peer evaluation. The main focus of the lessons was on the use of correct procedures to solve mathematical tasks. Errors made by pupils were not used as springboards for reflection. Also questions asked by pupils were not exploited to engage the class in critical thinking. Hence it may be said that although the learning environment did have the potential to stimulate the metacognitive potential of the pupils it was not harnessed. Cardelle-Elawar (1995) found that low achieving pupils benefited from metacognitive training.

In conclusion, we may claim that for the pupils in the LAPM study there is an apparent mismatch between pupils' preferred and teachers' mode of classroom instruction for the learning of mathematics. Teachers also appeared to have low expectations of their pupils in terms of what the pupils may be challenged with. They did not appear to engage their pupils in metacognitive activities, or challenge them with higher order thinking tasks, or pose them open questions during teaching and learning activities. Although teachers were very positive in terms of praising their pupils for correct responses and their attempts, they did not encourage pupils to clarify their doubts and thinking by asking questions or evaluating their peer's responses.

References

Angold, A. (1976). Diagnostic interviews with parents and children. In M. Rutter & E. A. Taylor (Eds.), *Child and adolescent psychiatry* (pp. 32-51). London: Blackwell Science Ltd.

Archer, A., & Isaacson, S. (1989). *Design and delivery of academic instruction*. Reston, VA: Council for exceptional children.

Cardelle-Elawar, M. (1995). Effects of metacognitive instructions on low achievers in mathematics problems. *Teaching & Teacher Education, 11*(1), 81-95.

Harris, C.A., Miller, S.P., & Mercer, C.D. (1995). Teaching initial multiplication skills to students with disabilities in general education classrooms. *Learning Disabilities Research and Practice*, *10*, 180-195.

Haylock, D. (1991). *Teaching mathematics to low attainers, 8-12*. London: Paul Chapman Publishing Ltd.

Jitendra, A.K., & Hoff, K. (1996). The effects of schema-based instruction on the mathematical word problem-solving performance of students with learning disabilities. *Journal of Learning Disabilities*, *29*, 422-431.

Jones, E.D., Wilson, R., & Bhojwani, S. (1997). Mathematics instruction for secondary students with learning disabilities. *Journal of Learning Disabilities*, *30*, 151-163.

Kroesbergen, E.H., & Van Luit, J.E.H. (2002). Teaching multiplication to low math performers: Guided versus structured instruction, *Instructional Science*, *30*, 361-378.

Reusser, K. (2000). Success and failure in school mathematics: effects of instruction and school environment. *European Child & Adolescent Psychiatry*, *9*, II/17 – II/26.

Van Luit, J.E.H. (1994). The effectiveness of structural and realistic arithmetic curricula in children with special needs. *European Journal of Special Needs Education*, *9*, 16-26.

Watson, A. (2001). Low attainers exhibiting higher order mathematical thinking. *Support for Learning, 16*(4), 179-183.

Wilson, R., Majsterek, D., & Simmons, D. (1996). The effects of computer-assisted versus teacher-directed instruction on the multiplication performance of elementary students with learning disabilities, *Journal of Learning Disabilities*, *29*, 382-390.

Zohar, A., & Dori, Y. J. (2003). Higher order thinking skills and low-achieving students: Are they mutually exclusive. *Journal of the Learning Sciences, 12*(2), 145-181.

Chapter 5

Recommendations for Teachers

Berinderjeet KAUR KOAY Phong Lee

FOONG Pui Yee CHANG Suo Hui

This chapter outlines the recommendations for teachers resulting from the research questions of the Low Attainers in Primary Mathematics (LAPM) study that were explored in this book. The recommendations are related to pupils' mathematics content knowledge, behaviours, affects and home backgrounds, and learning experiences.

1 Introduction

In this chapter we present the recommendations for teaching. They are based on the findings of research questions explored in chapters 2, 3 and 4 addressing the low attaining pupils' mathematics content knowledge, home related and self-related characteristics and learning experiences respectively.

2 Mathematics Content Knowledge

The main findings on the mathematical content knowledge of the low attainers and suggestions concerning the implications of the findings for the teaching of primary mathematics are given in this section. Both the findings and suggestions for teachers are organised according to the topics of the benchmark tests.

2.1 *Whole numbers*

A number, take eighty-two, can be represented in different ways: as a numeral 82, as number words "eighty two" or in concrete/pictorial presentation as

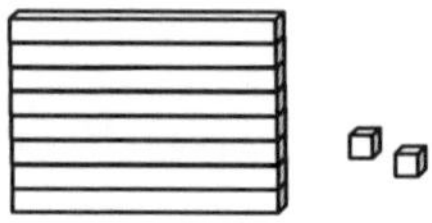

The benchmark test showed that some Primary 4 pupils still had difficulties with the different representations of numbers, including the conversion from one representation to another. In addition to the common spelling errors, many were confused between the use of terms such as hundreds and hundred. The latter is not a spelling error, teachers need to explain explicitly to the pupils the difference between number words and place value (e.g., 4 hundred and 4 hundreds). For common spelling errors in the written test, perhaps teachers could review the mark scheme so that pupils are not penalised for these errors.

Pupils had difficulties dealing with zero. They were not clear about its function as a placeholder. Some used it inappropriately as a place holder for the left most place and considered 013 as a three digit number. When teaching numeration, teachers should include examples that highlight the use of zero as placeholder and get the pupils to compare and contrast numbers such as 56, 506 and 560.

Pupils generally were quite proficient with the addition and subtraction of numbers, with / without renaming, but had difficulty with subtraction with multiple renaming involving zeros in the minuends. This shows that pupils had not mastered the procedure for renaming with deep understanding. They were not able to generalise the renaming process for subtraction when there is no zero in the minuend to subtraction involving zeros in the minuend. Hence, when teaching of subtraction with renaming, teachers could use manipulatives, whether concrete or virtual, to illustrate the process of "exchange 1 of this for 10 of that." to make sure that pupils learn the renaming procedure with relational understanding. They should help pupils see the connection between renaming involving no zeros in the minuends and renaming involving zeros in the minuend in subtraction.

Pupils were more proficient in recognising number patterns involving addition and / or subtraction rather than that involving multiplication and / or division. They did not perform as well when asked to explain the rules of the number patterns. Hence, choice of examples is important in the teaching of number patterns. Pupils should be exposed to different number patterns (involving arithmetic and geometric order) and patterns of different complexity. They should be given time and opportunity to communicate their thought processes and explain their solutions either using drawings and / or in words.

Many pupils had difficulty comprehending the word problems and consequently were not able to transform the information given in word problems meaningfully to relevant number sentences in order to solve the problems. Their performance was relatively poor with word problems involving multiplication and quotation (measurement) division. In most cases, they adopted the strategy of looking for key words to determine the operation required. Moreover, they believed that there could only be a solution to each problem. It is suggested that when teaching word problems, teachers should emphasise comprehension and representation of word problems, using think-out-loud to model the problem solving process. They should be aware of the different types of word problems and expose pupils to word problems involving similar operations in varied semantic structure. Moreover, class discussion on the alternative solutions to a problem should be included.

Besides the language difficulty and mathematical difficulty, the low attainers seemed to have limited working capacity that prevents them from holding amounts of information at one time. Consequently many of the errors they committed were due to partial recall or incorrect recall of information. Hence, exposing pupils to various memory strategies, showing them how to "cancel" to help free up working memory space (refer to last paragraph on number comparison), reduce the cognitive demand of the task, and organising information in word problems such as with a diagram / chart would help to improve pupils' performance.

2.2 *Fractions*

At this stage of learning the concept, the low attainers had the tendency to see a fraction not as a quantity but as two independent whole numbers separated by a short line, leading to add / subtract numerators and add / subtract denominators in the addition and subtraction of fractions. For those pupils who had partial recall of the procedure for addition / subtraction of fractions, they were likely to choose the larger denominator as the denominator for the answer. It is suggested that teachers should provide ample opportunities for pupils to distinguish equal parts and whole and examine the quotient relation between the numerator and denominator as well. This would help pupils to see a fraction as a number that can be indicated as a point on a number line. Perhaps curriculum planner and teachers could delay the instruction of addition and subtraction of fractions after pupils have conceptualised a fraction as a single quantity. In Singapore primary mathematics curriculum, fraction as a quotient is introduced only in Primary 5 after "fraction of a set" and the addition and subtraction of fractions. This could have explained why many pupils, particularly the low attainers, found it difficult to add and subtract unlike fractions correctly.

In fact, the geometric representation of "part of a whole" view of fraction used predominantly in the lower primary mathematics classroom might have reinforced the incorrect procedure in addition / subtraction of fractions. For many of these low attainers, they might look at the representation and add the number of shaded parts, the total number of parts and conclude that $\frac{2}{3}+\frac{1}{4}=\frac{3}{7}$.

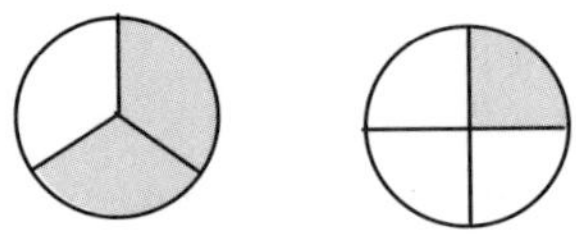

"Part of a whole" model of fraction is useful to help pupils established some basic ideas about fractions. However, curriculum planners need to consider whether fractions as quotients should be introduced earlier in the curriculum.

Many pupils were able to apply the "multiplying factor" procedure to find the missing numerator or denominator in a simple statement of equivalent fractions. However, they might have learnt the procedure for generating equivalent fractions with little understanding because they were not able to explain equivalent fractions in context, to apply the skills in comparing and ordering of fractions and in the addition and subtraction of unlike fractions. The "multiplying factor" procedure works because of the identity element for multiplication. Teachers need to explain this sophisticated idea to the pupils and not to focus just on the algorithmic approach of "multiplying the top and bottom by the same factor". The presentation of

$$\frac{2\times 2}{3\times 2}=\frac{4}{6} \text{ instead of } \frac{2}{3}\times\frac{2}{2}=\frac{4}{6}$$

reinforces the idea of fraction as two different numbers in pupils. In addition, teachers could link the equivalent fractions concept taught in class with the children's intuitions on equal sharing, and should not limit their instruction to the partition context. For example, 3 children sharing 2 pizzas is the same as 6 children sharing 4 pizzas.

Many low attainers were successful at addition / subtraction of like fractions but performed poorly at addition / subtraction of unlike fractions. Equivalent fractions concept is one of the prerequisites for the addition and subtraction of unlike fractions. How it is applied in the algorithm for the addition and subtraction of unlike fractions should be explained explicitly to the pupils. Helping pupils to establish the connection between the algorithms for the unlike fractions and that for the like fractions would help to lighten the task demand on the working memory of the pupils.

Similar to the findings for the whole numbers, pupils had more difficulty with compare word problems than join / separate problems involving fractions. They tend to focus on the comparative words such as "more" and "less" and ignore the "unknown". As a result, incorrect operation was used to solve the problem. This shows that the teaching of the "key word" strategy in solving word problems involving whole numbers could inhibit pupils' problem solving ability. Pupils should be encouraged to read and understand the problem statements and make sense of their actions and solutions.

To improve pupils' ability in solving word problems, the curriculum should include various types of word problems: join / separate, part-part-whole and compare. Teacher should select with care the examples used during instructions and for pupil practice so that incorrect intuition and informal conceptions of fraction are revealed, monitored and remediated.

2.3 *Measurement*

The majority of the Primary 4 low attainers had no difficulty using the standard ruler to measure length by aligning the end-point of the ruler with one end of the line and read off the number of centimetres on the ruler. However, many pupils had not understood the idea of iteration of standard units (cm) and failed to realise that the length of the line being measured is the sum of the number of units in the part of the ruler that correspond to the line. Hence any point on the ruler can be the "starting point". Teachers should be aware of the difficulty pupils have with the iteration and demonstrate how measurement is based on the one-to-one correspondence between the number of iterated units (for example, paper clips in the figure below) and the imaginary units on the object.

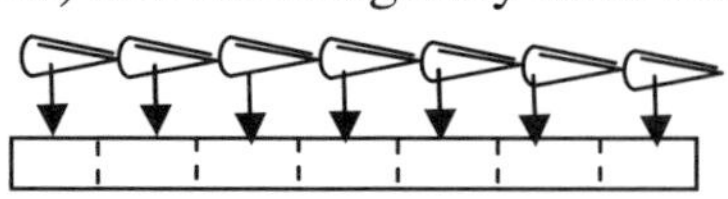

Before the introduction of the standard ruler, teachers may get pupils to construct their non-standard unit rulers (e.g., putting a string through straws of equal length). As the pupils use such ruler to measure the length of an object, their focus would be on the iterated units rather than the numbers as marked on the standard ruler. Later, teachers can use a broken ruler to help pupils realise that any point on a ruler can be the "starting point".

The Benchmark test on measurement also showed that many low attainers had not understood the inverse relation between the size of the unit and the number of units required to measure. Apparently, pupils have not had sufficient practice on the use of different non-standard units and of different sizes to measure an object. Teachers should use the experiences to lead pupils to conclude that you need shorter / smaller units than longer / larger units to measure an object.

Conversion of units was not easy for the low attainers as they were confused with the equivalence. Often, they were aware that the relationship between two units is in the power of 10 but they were not sure which. For many of these pupils, visualisation may assist them to recall the relationship. Teacher should use concrete presentation to illustrate the relative size of measurement units when possible so that pupils have a mental image of the comparison. For example, place a centimetre cube next to a metre ruler and get pupils to "see" the relative sizes of centimetre and meter. Later, teachers could then encourage pupils to recall their visual images of the comparison and reflect on whether their conversions make sense.

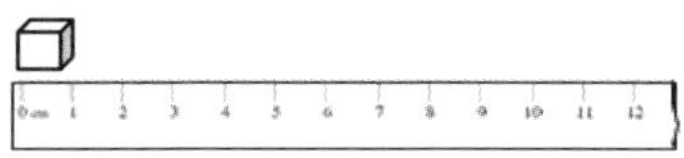

Many low attainers could not distinguish the concept of area from the concept of perimeter, while others did not understand the measuring units for area. The pupils were more successful with rectilinear figures. When the figure was not rectilinear, they either counted the diagonal of the small square unit as 1 unit or considered the area of the right angled isosceles triangle as 1 unit^2 instead of $\frac{1}{2}$ unit^2. It is suggested that teachers provide plenty of well-designed hands-on learning activities for the pupils to measure round a shape and to use unit squares to cover the shape, explain clearly the difference between perimeter and area. To address the problem of counting the units of shaded area, teachers may perhaps get pupils to compare figures such as these to convince the pupils that their method of counting is incorrect.

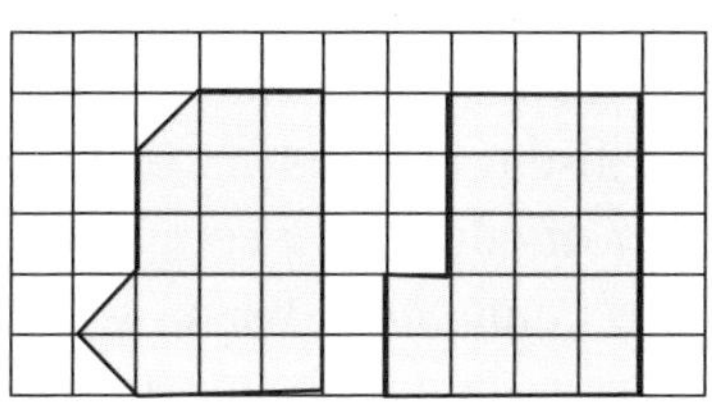

Pupils' erroneous way of counting would lead them to conclude that both figures had the same area but what they see would not support their conclusion.

Money is an important part of their everyday life but many Primary 4 low attainers had difficulty with the decimal representation of money. They did not understand the decimal point in the representation and the meaning of equivalence. They wrote \$37.20 = 20¢. In real life, no Primary 4 child would exchange \$37.20 for 20 cents. Here, teachers must emphasise the meaning of equivalence and the meaning of the "=" sign. They need to design their curriculum to help pupils make the connections between mathematics learnt in the classroom with that outside the classroom.

As reported under the sections on whole numbers and fractions, solving word problems in measurement was not easy for the low attainers. Language comprehension seemed to be the major learning difficulty. The structure of the word problems also affected pupils' performance. Measurement provides a context for arithmetic word problems. Teacher should provide step-by-step guidance, avoid key-word strategy that tends to mislead pupils to use incorrect operation in indirect language problems. Give pupils opportunities to reflect on the problem solved, discuss how the structure of one problem is different from the other, and how visual representation such as the model may help them arrive at the correct number sentence. Pupils should be encouraged to check that their answer to the word problem makes sense.

2.4 *Geometry*

Some pupils considered the edges of a composite shape as the sum of the edges of the basic shapes. To improve pupils' performance on this type of items, let pupils create a composite shape from basic shapes, trace the outline of the shape and discuss the edges of the composite figure. Let pupils see that the number of edges of a composite figure is less than the sum of the edges of the component basic shapes. Teacher can also reverse the process, and have the pupils "decompose" a given shape into different basic shapes, a prerequisite for the area of the composite figure.

Many pupils had difficulty with identification of perpendicular lines that were not horizontal lines or vertical lines. They were successful in identifying parallel lines. Perpendicularity is a basic geometric relation. The main situations giving sense of this concept before the introduction of vertical and horizontal lines in Primary 4 are the recognition of certain basic geometrical objects and a right angle in a plane figure and square grid. To improve the success rate in identifying perpendicular lines, instruction on perpendicular lines should include examples and non-examples in different orientations to help pupils focus on the invariant features. It should also include direct instruction about the means to test for perpendicularity.

Angle is an abstract concept, a difficult concept for all pupils, particularly the low attainers. One of the learning difficulties seemed to be the idea of equivalency. About a third of the pupils chose only one angle when they were asked to select two equal angles. However, they were more successful when asked to identify angle greater / smaller than a right angle. Perhaps, as teachers teach pupils to compare angles, they should also include equal angles. Comparing should not be limited to cases involving greater and smaller. Moreover, teacher should teach these low attainers how they can use the right corners of familiar objects such as corner of an A4 paper as a benchmark to identify obtuse and acute angle, if they are not sure.

Completing a shape pattern involving more than one attribute was not an easy task for the low attainers. Teacher should expose pupils to various types of pattern, starting with patterns involving one attribute. Attribute blocks can be used in the related hand-on activities. For a complex shape pattern, teachers should explain explicitly the strategy of first identifies the attributes in the pattern and then considering one attribute at a time. e.g., in the following pattern

First, consider the shape; the missing shape would be a square in this AB pattern. Then consider the colour, it would be black in the AAB pattern. So the missing item would be a black square. It is advisable to give short but regular practice to help the low attainers master the strategy.

2.5 *Data analysis*

Pupils were better at reading the data from the graph than reading between the data. These low attainers did have the ability to learn, so instruction on graph comprehension should not be limited to reading from the data, it should also include questions related to higher levels of graph comprehension.

Pupils had difficulties interpreting the scales in both the picture graphs and the bar graphs. Many of them seemed not to consider scale as an important element of the graph. To make pupils aware of the use of scales, teachers may present a scenario when the use of scale is essential. For example, for picture graph, let pupils think about what to do when there are insufficient icons available to construct the graph, and discuss the commonalities and the structural relationships between graphs with scale and those without scale. Pupils need to develop a deep understanding of scales in picture graphs and scale markings in the bar graphs.

The low attainers were easily distracted by the salient perceptual features of the graph. For example, when given a bar graph with five categories, one of which has a value of zero, pupils would state that there were four categories of data in the graph. Teachers hence should focus pupils' attention on the framework of the graph and its referents rather than the visual appearance of the pictures or bars. For example, what information is being presented in the graph and so what would be an appropriate title for the graph, how the axes are labelled and what is the scale used.

For questions involving reading between the data, many low attainers used indiscriminately the strategy of "correctly identified values and added" when they noted the word "more" in the question. To promote graph comprehension and mathematical communication, teachers could provide a graph and have pupils tell a story based on the information given in the graph. Pupils could then be encouraged to use appropriate mathematical language to describe the relationships represented.

The low attainers were less successful in completing a graph than reading the data from the graph. "Completing a bar graph from given data" is one of the instructional objectives for Primary 3 mathematics. However, graphing activities in lower primary mathematics are often limited to questions that involve level one of graph comprehension. All pupils including the low attainers should have experience in problem solving that requires them to present the answer in the graph. They should then reflect on the reasonableness of their responses.

3 Behaviours, Affects and Home Backgrounds

The findings on low attainers' non-cognitive domains help in some ways to interpret their underachievement in mathematics. These pupils' affects in learning and their behaviours in the classroom can be seen as a barometer of the quality of teaching and learning that these low attainers are experiencing in the day-to-day mathematics lessons. It is not reasonable to infer that pupils bring their attitudes to class that would influence their learning in normal circumstances. For primary school pupils, most likely it is the learning experiences in school that influence their attitudes. Knowing the kind of home backgrounds that the low attaining pupils come from would only make teachers more aware of the educational resources or the lack of which these pupils can or cannot do the work that teachers assigned them for homework. From the expressed feelings of the low attainers about their self-concepts as learners of mathematics and their feelings towards learning the subject, they point to the bad effects of too many tests and assessment in our education system. Pupils see grades from their mathematics tests as information about their ability and therefore judge themselves highly if they did well or lowly if they did not pass. The poor results that the low attainers obtained so frequently are sending the message that they are not good in mathematics and that they are failures. This kind of effect is detrimental to future learning as they set off tension and anxiety in the pupils who already are anticipating difficulties in mathematics.

The content of mathematics gets more abstract and further away from pupils' everyday experiences as they move up the level. In the local primary mathematics curriculum, there are many abstract structures in topics such as the four operations of fractions and even the use of the model method that is meant to help pupils solve complex relationship word problems. In this study many of the low attainers have identified these topics as difficult for them. If they are not taught well to low attainers with their characteristics behaviours and affects, then it is no surprise that these pupils found it difficult to understand and thereby do not like mathematics. For the perpetual failures it is also no surprise that misbehaviours like inattentiveness and disruptions will occur when they are bored and confused. Knowing that circumstances surrounding the learning experience influence the learning attitudes, the following are some suggestions for teachers.

3.1 *Use the C-P-A approach consistently and thoughtfully*

Make mathematics fun through the use of concrete models and activity-based lessons. Presentation of the content must be given careful consideration and to be covered in ample depth and over adequate time to allow assimilation by the pupils who may learn at a slower pace. Demystify mathematics by creating appropriate classroom environment where students perceive mathematics as sense-making activity and be able to see connections across topics and in real-life context. Reinforcement through practice can be made interesting through games and link to materials used in the C-P-A approach. Pupils are more interested in the manipulative activities than always the paper-pencil worksheet. Teachers should develop a reasonable level of tolerance towards pupils who tend to be fidgety. Give pupils a few seconds to "stand up and stretch" time during the lesson.

3.2 *Model problem-solving strategies rather than model-answers or techniques*

Let pupils present suggestions, try their ideas and let them see why they do or do not work. Allow pupils to participate interactively through cooperative small group discussion and problem solving. Teach pupils to be mathematicians at their own level rather than teaching them to just learn to perform routine operations. Solve problem together with pupils and discuss their common error and difficulties. Do not assume the model-method itself is a help for pupils to solve word problems. Transformation from text to the model drawing is a higher order thinking skill for the low attaining pupils with limited language facility. Teacher should model the Polya stages of problem solving with due consideration of comprehension of terms and semantics in word problem, identification of the generator of the unit to represent the elements and careful labelling of all the known and unknowns in the problem situation.

3.3 *Give pupils opportunity for metacognition and communication in mathematics through journal writing*

Writing about how they feel about a lesson, what they have understood and how they solve a problem will provide teacher with deeper understanding of their pupils' thoughts and feelings in and out of the classroom. Provide positive feedback that focuses on what pupils do right and treat mistakes as a normal part of learning. Talk with pupils about their anxiety or mathphobia and accord recognition to their problems. Encourage pupils to share their journal writings with the class, in this way the pupils would feel "less stupid" and not alone in their feelings.

To facilitate teachers in implementing innovative teaching strategies and alternative forms of assessment to meet the needs of low attainers in mixed-ability classes of usually 30-40 pupils, the support of school is important. Consideration to class size and provision of teacher aide in a mixed-ability classroom can help a teacher in conducting differentiated instruction that incorporate more hands-on group activity-based lessons for pupils to develop conceptual understanding and communication in mathematics. School could re-look the mathematics syllabus for the lower primary levels in term of content re-designation or reduction for those pupils who would need more curriculum time in their multi-level cognitive development of the various topics in the syllabus. By the time a cohort of low attainers reached Primary 5 where they could be streamed into the Foundation Course, it will be too much and too late to fill the large gap of learning. There are low attainers who failed to achieve the learning outcomes set out in the syllabus since Primary 1. Regardless, teachers often rushed to finish the syllabus for a cohort in order to meet the required formal assessments such as continual assessments (CA) and semestral assessment (SA). This in turn generated a cumulative and increasing number of failures as they move up the level. Other policies such as assessment and banding of pupils need to be re-examined for better learning and teaching, if not to increase the pool of low attainers.

4 Learning Experiences

In this section we present our recommendations in areas of concern that have resulted from our findings on the learning experiences of low attainers in primary mathematics. The recommendations mainly are for mathematics instruction, and mathematical tasks and pupil talk.

4.1 *Mathematics instruction*

From the lessons observed, it was apparent that teacher-led whole class instruction appeared to be the type of instruction almost all the teachers in the study adopted most of the time. Their classroom instruction appeared to have elements of direct instruction – such as step by step development of the concepts and skills but lacked the requisite of mastery by pupils at each step. Research has shown that teacher-led direct instruction is effective with pupils having difficulties in mathematics (Harris, Miller, & Mercer, 1995; Jitendra & Hoff, 1996; Van Luit, 1994; Wilson, Majsterek, & Simmons, 1996). Therefore we suggest that teachers fine tune their teaching plans and execute direct instruction during their lessons. Direct instruction is a teacher-centred instructional approach that is systematic and follows a sequence of steps, generally stating the objective of the lesson, reviewing skills necessary for new information, presenting new information, questioning pupils, providing group instruction and independent practice, assessing performance and giving more practice (Swanson, 2001). A significant characteristic of this mode of instruction is the requirement of mastery by the pupils at each step (Kroesbergen & Van Luit, 2003).

Data from the Pupil Maths Learning Questionnaire revealed that a high proportion of pupils expressed a desire to work in groups and learn mathematics by doing activities. These findings signal a need for teachers to consider incorporating group work in their lessons. We would like to suggest that teachers punctuate their classroom instruction with groupwork at appropriate junctions. They must give carefully plan and select activities for group work so that pupil learning is the focus of such sessions.

It is apparent from the pupil interview data that pupils did not like to work on their own as they lacked the confidence to attempt tasks without the support of their peers and teachers. Such findings are important for teachers as they guide them in planning fruitful lessons. We encourage teachers to seek the perspectives of their pupils occasionally about their preferences and difficulties so that the gap between teaching and learning may be optimised.

4.2 *Mathematical tasks and pupil talk*

From the lessons observed in the study, it was apparent that teachers did not emphasise higher order thinking in their classrooms. Watson (2001) and Zohar and Dori (2003), in their work with students have shown that pupils can benefit from challenging tasks, provided that they are appropriately structured. We like to encourage teachers of low attainers to integrate the common mistakes their pupils make into their classroom activities. This is possible through a strategy called "What's wrong?" introduced by Kaur and Yeap (2009) to primary school teachers across all schools in Singapore.

In "What's wrong?" kind of tasks pupils are presented with a problem and an erroneous solution that may be conceptual or computational. The pupil has to recognise the error, correct it and then explain what was wrong, why it was wrong and what was done to correct the error (Krulik & Rudnick, 1999). Such tasks demand higher order thinking, namely critical thinking. Pupils may be asked to complete the task in small groups. The teacher must ensure that pupils are engaged in class discussion after completing the task so that they get the opportunity to see ways of solving problems that differ from their own. Furthermore, these discussions often lead to deeper mathematical understanding (Krulik & Rudnick, 2001). Teachers are in a good position to craft tasks like this as they are constantly exposed to errors pupils make in class and in their written assignments. Figure 1 shows one such task.

What's wrong?	
Mrs Lim had 12 boxes of apples. Each box had 48 apples. She threw away 32 rotten apples. How many apples had she left?	
Peter's solution: 48 – 32 = 16 16 x 12 = 192 Ans: 192 apples left.	Rani's solution: 12 x 48 = 576 567 – 32 = 535 Ans: Mrs Lim had 535 apples left
There is something wrong with Peter's and Rani's solutions? Explain their mistakes. Show how you would find the solution to the problem.	

Figure 1. Example of a "What's wrong?" task

When teachers implement a task like the one shown in Figure 1, they are not only providing their pupils with an opportunity to engage in higher order thinking but also to reason and communicate their thoughts during the whole class discussion session that follows group activity. Such tasks will also provide teachers with ample opportunities to engage pupils in reflection through the open questions that they may pose the class.

The tasks used in the lessons observed were found to be routine and repetitive. For pupils to develop procedural fluency a certain number of repetitive tasks are necessary to be part of a lesson. However, extended periods of repetitive tasks may not be engaging. We suggest that teachers look for different approaches of presenting the same content and skills for practice. One good strategy that teachers may like to explore would be "What's the question?" (Kaur & Yeap, 2009). In "What's the question?" kind of tasks pupils are presented with a context and data but with question/s missing. Figure 2 shows one such task.

For a task like the one shown in Figure 2, pupils may be asked to work in groups and discuss their answers to parts i and ii. They may also be asked to write a question for their friend. Such a task provides opportunities for significant pupil-pupil talk. The teacher can also help to maintain the cognitive demand of the task, by asking open questions, such as "why do you think so?", or "is this the only way to do it?", etc.

We like to discourage teachers from asking "any questions" as a general question to which often the response is silence. Following group work by pupils, the teacher may ask the groups to present their questions and thinking to the class. Pupils may also be encouraged to evaluate their peers' responses.

Funland Circus

Type of ticket	Cost of ticket	
	weekday	weekend
Adult	$15.00	$30.00
child	$ 8.50	$15.00

i) What's the question if the answer is $23.50?

ii) What's the question if the answer is $45?

Figure 2. Example of "What's the question" task

It was apparent in the lessons observed that teacher talk dominated during the instructional period. This apparently was a consequence of the instructional method adopted by the teachers which was mainly teacher demonstration, seatwork and review of pupil work. We strongly urge teachers to provide pupils with opportunities to talk and clarify their thinking. As we have shown, this is possible with mathematical tasks like the ones shown in Figures 1 and 2. Such tasks are easily crafted from pupil errors and textbook questions.

Finally, we like to encourage teachers to leverage on their classroom environments which were warm and non-threatening to motivate pupils to ask questions and freely share with others their thoughts. Use pupil errors as springboards for learning by asking open questions, engaging pupils in reflection and evaluation. We strongly urge teachers to elevate their expectations of low attainers under their charge and help them achieve their potential.

References

Harris, C.A., Miller, S.P., & Mercer, C.D. (1995). Teaching initial multiplication skills to students with disabilities in general education classrooms. *Learning Disabilities Research and Practice*, *10*, 180-195.

Jitendra, A.K., & Hoff, K. (1996). The effects of schema-based instruction on the mathematical word problem-solving performance of students with learning disabilities. *Journal of Learning Disabilities*, *29*, 422-431.

Kaur, B., & Yeap, B.H. (2009). *Pathways to reasoning and communication in the primary school mathematics classroom.* Singapore: National Institute of Education.

Kroesbergen, E.H., & Van Luit, J.E.H. (2003). Mathematical interventions for children with special educational needs. *Remedial and Special Education*, *24*, 97-114.

Krulik, S., & Rudnick, J.A. (1999). Innovative tasks to improve critical and creative thinking skills. In L. Stiff (Ed.), *Developing mathematical reasoning in grades K-12* (pp. 138-145). Reston, VA: National Council of Teachers of Mathematics.

Krulik, S., & Rudnick, J.A. (2001). *Roads to reasoning — Developing thinking skills through problem solving* [Grades 1 – 8]. Chicago, IL: Wright Group McGraw-Hill.

Swanson, H.L. (2001). Searching for the best model for instructing students with learning disabilities. *Focus on Exceptional Children*, *34*, 1-15.

Van Luit, J.E.H. (1994). The effectiveness of structural and realistic arithmetic curricula in children with special needs. *European Journal of Special Needs Education*, *9*, 16-26.

Watson, A. (2001). Low attainers exhibiting higher order mathematical thinking. *Support for Learning, 16*(4), 179-183.

Zohar, A., & Dori, Y. J. (2003). Higher order thinking skills and low-achieving students: Are they mutually exclusive. *Journal of the Learning Sciences, 12*(2), 145-181.

Appendices

Appendix A

School Management Questionnaire

School: ____

Name 1: ______________________ Appointment: ______________

Name 2: ______________________ Appointment: ______________

Name 3: ______________________ Appointment: ______________

Dear Participants
The purpose of this questionnaire is to gather baseline information about the practices used by primary schools in Singapore in identifying, organising and motivating the group of low attaining pupils, particularly in Mathematics. To help us in our study of low attaining pupils, we would be grateful if you could furnish us with the information pertaining to your school practices.

How does your school identify/screen low attainers in mathematics across the levels (P1–P5)? Please describe your practices, both at the departmental and classroom levels.

Level	Practices
P1	
P2	
P3	
P4	
P5	

1. How does your school classify / group low attaining pupils across the levels?

____

____

2. What strategies does your school use in organising instructional programmes for supporting and motivating low attaining pupils across
   the grade levels:
   a. During curriculum time

   ____

   ____

   b. Out of curriculum time

   ____

   ____

3. How are the practices of identifying low attainers in English Language integrated with the identification of low attainment in mathematics?

____

____

Thank you for your participation

Appendix B

Teacher Profile Questionnaire

Name ______________________ School ______________________

Gender ______________________ Email ______________________

Age ______________________

1 Highest level of academic qualification

☐ 'A' Level ☐ Diploma ☐ Degree

☐ Masters ☐ Others ______________ (Please specify)

2 Highest level of professional qualification

☐ Cert. Ed ☐ Dip. Ed./ Dip.PE ☐ PGDE

☐ B. Ed. ☐ M. Ed ☐ Others ______________ (Please specify)

3 Highest academic qualification in Mathematics

☐ 'O' level ☐ 'A' level ☐ Polytechnic level

☐ Undergraduate level ☐ Master

4 Current appointment

☐ Teacher

☐ Subject / Level Head Subject /Level ______________________

☐ HOD Department ______________________

☐ Adjunct ☐ Relief

5 How long have you taught primary mathematics in school?

☐ less than 1 year ☐ 7 to 10 years

☐ 1 to 3 years ☐ More than 10 years

☐ 4 to 6 years

6 How many years have you taught mathematics to Primary 3/4 pupils?
About ____________ years.

7 Have you been assigned to teach low ability class/group (mathematics) before? If yes, for how many years?
☐ No ☐ Yes, about _______ years

8 Have you been trained for the LSM programme? (LSM- Learning Support in Mathematics)
☐ No ☐ Yes

9 Have you taught in LSM classes before? If yes, for how many years?
☐ No ☐ Yes, about _______ years

10 Have you attended any in-service training related to teaching low attaining pupils? If so, please specify:

____

____

____

Thank you for your participation

Appendix C

Teacher Interview Prompts

No.	Prompt	Category of data
1	Please state your name, school and appointment.	General information
2	What is your understanding of 'low attainers'?	General information
3	What are the general characteristics you observe in your pupils who are low attaining in Mathematics?	General information
4	What do you think are the main causes of Math learning difficulties for low attaining pupils?	General information
5	What do you know about the practices used by your school to identify low attaining pupils?	School
6	Are there any specific practices used by your school to identify low attaining pupils based on each subject? If so, please explain.	School
7	What are the practices in your school for the early screening of low attaining pupils?	School
8	What are the practices used by your school to organise low attaining pupils?	School
9	What are the practices used by your school to motivate low attaining pupils?	School

No.	Prompt	Category of data
10	Are there any practices other than the ones set by the school, which you follow to identify low attaining pupils in mathematics?	Teacher
11	Are there any practices other than the ones set by the school, which you follow to organise low attaining pupils in mathematics?	Teacher
12	Are there any practices other than the ones set by the school, which you follow to motivate low attaining pupils in mathematics?	Teacher
13	What do you think (feel) are the limitations of the current practices in your school in identifying and organising low attaining pupils? What recommendations would you like to make to bring about changes to these practices?	Teacher

Appendix D1

Whole Numbers

The tests (Parts I, II, and III) are based on the following content from the primary school mathematics syllabus (Ministry of Education, 2006) in Singapore.

Primary 1	Primary 2	Primary 3
▪ numbers up to 100 ▪ addition & subtraction ▪ mental calculation multiplication & division	▪ numbers up to 1000 ▪ addition & subtraction ▪ multiplication & division ▪ mental calculation	▪ numbers up to 10 000 - read and write numbers - compare and order numbers - odd and even numbers - number patterns ▪ the four operations - add and subtract numbers up to 4 digits - multiplications and division of numbers up to 3 digits by 1 digit - division with remainder - solve up to 2-step word problems ▪ mental computation - addition and subtraction involving 2-digit numbers - multiplication and division within the multiplication table.

Part I

1 Write 5802 in words.

[ ]

2 Write eight thousand and thirty two in numerals. [ ]

3 Look at the pictures and answer the following questions.
The animals are queuing for food.

a) The animal at the end of the queue is the [ ]

b) The hen comes just before the [ ]

c) The bird is the [ ] animal in the queue

4 In 9853, the 8 stands for 8 hundreds.

a) In 9083, the 8 stands for [ ]

b) In 8930, the 8 stands for [ ]

5 a) In 8906, the 9 is in the [ ] place.

b) In 7004, the 7 is in the [ ] place.

6 a) 879 —Add 100→ [ ]

b) 4032 —Add 100→ [ ]

7 a) 5249 —Subtract 1000→ [ ]

b) 1876 —Subtract 1000→ [ ]

8 Form the smallest **4-digit** number with these digits.

0	4	5	9

9 Form the largest **4-digit** number with these digits.

1	0	8	3

10 Write a number that is one more than 9899.

11 Write a number that is greater than 5267 but less than 5367.

12 Arrange these numbers from the smallest to the largest.

6402 6240 4062 4602

13 Arrange these numbers from the largest to the smallest.

398 908 389 983

14 Circle all the even numbers.

230 189 457 718

15 Find the missing numbers in the number patterns. What is the rule? The first one is done for you.

Example: 107 110 113 116 119

Rule: Add 3 to the number before.

a) 302, 297, 292, 287, ______

Rule: ______________________

b) 2, 4, 8, 32

Rule: ______________________________

c) 78, [ ], 278, 378, [ ]

Rule: ______________________________

16 Circle **one** number that has a value nearest to the number on the left.

a) 644 ⟶ 650 600 700 640

b) 2097 ⟶ 2000 2100 2090 2095

17 Fill in the missing numbers in the blanks

a)

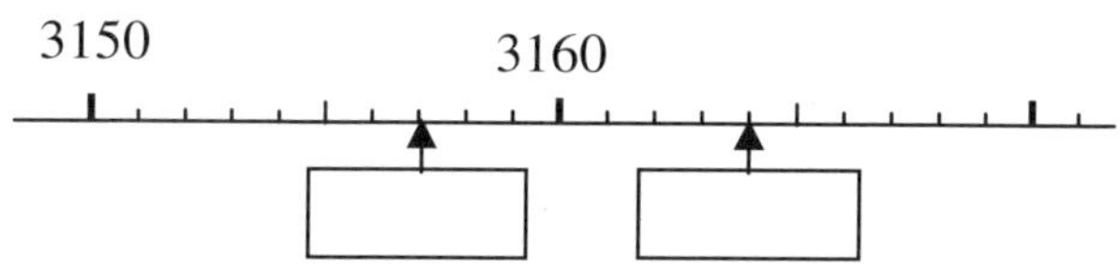

b)

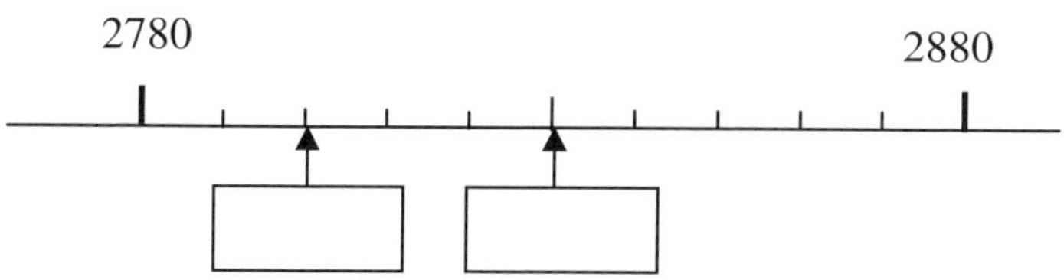

Part II

18 Read each question carefully.
Circle the operation you will use.

Example: Jane has 160 stamps. Her father gives her 20 more stamps. How many stamps does Jane have now?	(160 + 20) (circled) 160 x 20	160 – 20 160 ÷ 20
a) Ali had some guppies. He bought another 10 guppies. He has 280 guppies now. How many guppies did he have at first?	280 + 10 280 x 10	280 – 10 280 ÷ 10

b)	An apple costs 45¢. What should the price of 5 apples be?	45 + 5 45 x 5	45 – 5 45 ÷ 5
c)	A bar of chocolate can be broken into 18 similar squares. There are 3 squares in a row. How many rows are there in the chocolate bar?	18 + 3 18 x 3	18 – 3 18 ÷ 3
d)	Mr Lim walks 12 km home from work. He stops at a shop after walking 4 km. How much further has Mr Lim to go?	12 + 4 12 x 4	12 – 4 12 ÷ 4
e)	Salma packed 36 cookies in some bags. She put 4 cookies in each bag. How many bags did she have?	36 + 4 36 x 4	36 – 4 36 ÷ 4
f)	Ali has $45 and Meng Ho has $5. How much more does Ali have than Meng Ho?	45 + 5 45 x 5	45 – 5 45 ÷ 5
g)	An oven tray can hold 6 cakes. Mrs Tan fills 30 trays. How many cakes will Mrs Tan bake?	30 + 6 30 x 6	30 - 6 30 ÷ 6

19 Divide 122 by 8.

What is the quotient? ☐

What is the remainder? ☐

20 Fill in the missing numbers.

a) 127 + 42 = ☐

b) 423 + 261 = ☐

c) 264 + 126 = ☐

d) 378 + 145 = ☐

e) 247 + 153 = ☐

f) 607 + 393 = ☐

21 Fill in the missing numbers.

a) 286 - 25 = ☐

b) 647 - 326 = ☐

c) 739 - 497 = ☐

d) 456 - 278 = ☐

e) 806 - 439 = ☐

f) 600 - 437 = ☐

22 Fill in the missing numbers.

a) 4 x 6 = ☐

b) 5 x 8 = ☐

c) 9 x 7 = ☐

d) 6 x 8 = ☐

e) ☐ x 7 = 14

f) ☐ x 4 = 28

g) 5 x ☐ = 30

h) 7 x ☐ = 56

Part III

23 30 cards are shared out equally among 5 children.
How many cards does each child have?

☐ cards

24 Jane has $100. She earns another $50 more and spends $80.
How much does she have now?

$ ☐

25 Each ticket costs $6.
How many tickets can be bought with $48?

☐ tickets

26 I buy two books costing $14 and $20.
What is my change from $50?

$

27 Siti packs some tarts in 9 boxes.
She packs 4 fruit tarts and 5 egg tarts in each box.
How many tarts does she pack altogether?

tarts

28 Sally wants to buy a piano that costs $3010.
She saves $200 each week for 3 weeks.
How much more money does she need to save?

$

29 Rohani has 50 envelopes and 220 cards.
She puts 5 cards in each envelope.
How many envelopes are left unused?

envelopes

30 Mr Ali had 5000 bottles of cooking oil in his store room.
He sold 895 bottles in the first month and 1042 bottles in the second month. How many bottles of cooking oil had he left?

bottles

31 400 concert tickets were sold on Saturday.
163 more concert tickets were sold on Sunday than on Saturday.
How many concert tickets were sold on both days?

tickets

Appendix D2

Fractions

This test is based on the following content from the primary school mathematics syllabus (Ministry of Education, 2006) in Singapore.

Primary 1	Primary 2	Primary 3
	▪ fraction of a whole ▪ addition & subtraction	▪ equivalent fractions ▪ addition & subtraction of 2 related fractions within one whole

1 The circle is cut along the line.

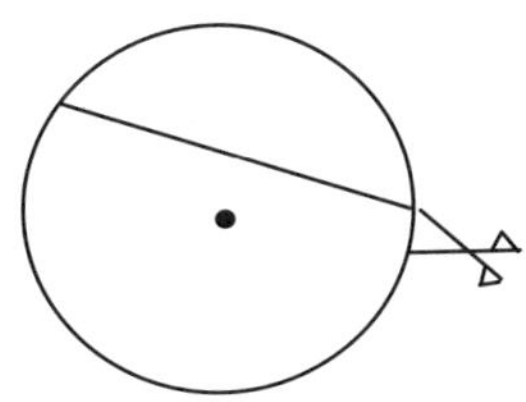

a) Has the circle been cut into 2 halves? ______________

b) Why? You may use diagram to explain.

____

2 Share this bar of chocolate equally among 5 children.

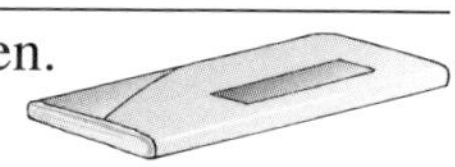

What fraction of the chocolate bar does each child get?

3 Shade in two-thirds of this rectangle.

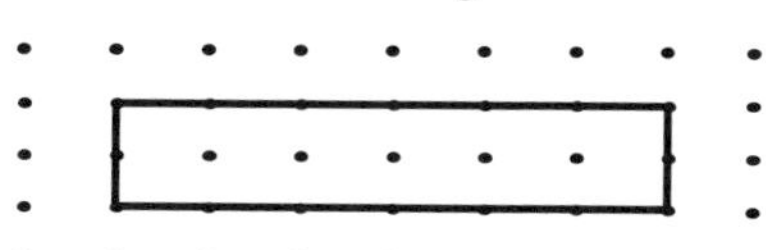

4 What fraction of each of the shapes below is shaded?

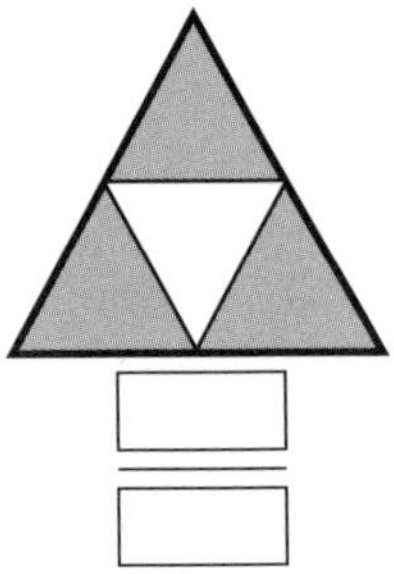

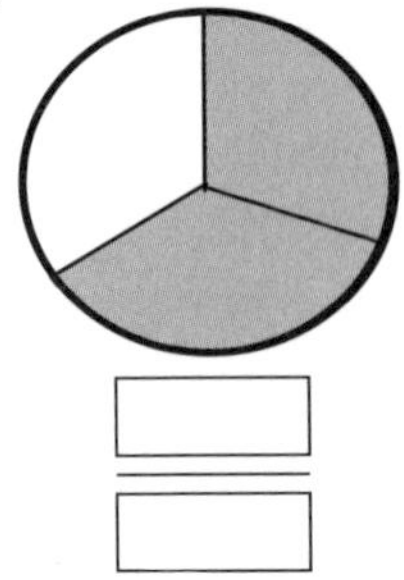

5 Circle the larger fraction in each of the pairs below.
The first is done for you.

a) $\frac{1}{4}$ $\frac{3}{4}$ (circled)

b) $\frac{3}{7}$ $\frac{5}{7}$

c) $\frac{3}{5}$ $\frac{3}{4}$

d) $\frac{2}{5}$ $\frac{3}{10}$

6 Put these fractions in order of size, from the largest to the smallest.

$\frac{1}{7}$ $\frac{1}{4}$ $\frac{1}{12}$

7 Put a tick under the shape(s) that is shaded half.

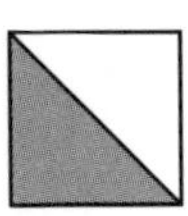

( )

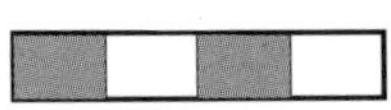

( )

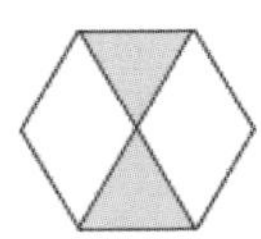

( )

8 Meng and Ahmad each have a bar of chocolate of the same size.
Meng breaks his into 6 equal pieces and eats 4 of them.
Ahmad breaks his into 3 equal pieces and eats 2 of them.

a) Who eats more chocolate? ____________

b) Why do you think so? You may use diagram to explain.

____

9 Fill in the missing numbers:

a) $\frac{1}{3} = \frac{2}{\square}$ b) $\frac{6}{8} = \frac{3}{\square}$

c) $\frac{6}{12} = \frac{\square}{4}$ d) $\frac{2}{5} = \frac{\square}{10}$

---

10 Add

a) $\frac{2}{7} + \frac{3}{7} =$ b) $\frac{7}{10} + \frac{1}{5} =$

c) $\frac{1}{3} + \frac{1}{9} =$ d) $\frac{3}{8} + \frac{2}{4} =$

---

11 Subtract

a) $\frac{7}{8} - \frac{2}{8} =$ b) $\frac{5}{6} - \frac{1}{2} =$

c) $1 - \frac{3}{5} =$ d) $\frac{2}{3} - \frac{1}{6} =$

---

12 Maylin used $\frac{3}{5}$ kg of flour for bread and $\frac{3}{10}$ kg of flour for cakes.

How much flour did she use altogether?

[ ] kg

---

13 Jane has $\frac{1}{6}$ m of ribbon.

Nurul has $\frac{7}{12}$ m of ribbon.

a) Whose ribbon is longer?
b) How much longer?

a) [ ]

b) [ ] m

---

14 There was some juice in the glass.

Devi drank $\frac{1}{4}\,l$ of the juice from the glass.

There is $\frac{3}{8}\,l$ of juice left in the glass.

How much juice was there in the glass at first?

	l

Appendix D3

Measurement

The tests (Parts I and II) are based on the following content from the primary school mathematics syllabus (Ministry of Education, 2006) in Singapore.

Primary 1	Primary 2	Primary 3
▪ length & mass (non – standard units) ▪ time (to the hour/ half hour) ▪ money	▪ length, mass & volume (standard units) ▪ time (minutes) ▪ money (decimal notation)	▪ length, mass & volume ▪ time ▪ money • area & perimeter

Part I

1 Compare the lengths of the lines AB, CD, EF and GH drawn on the grid paper below.

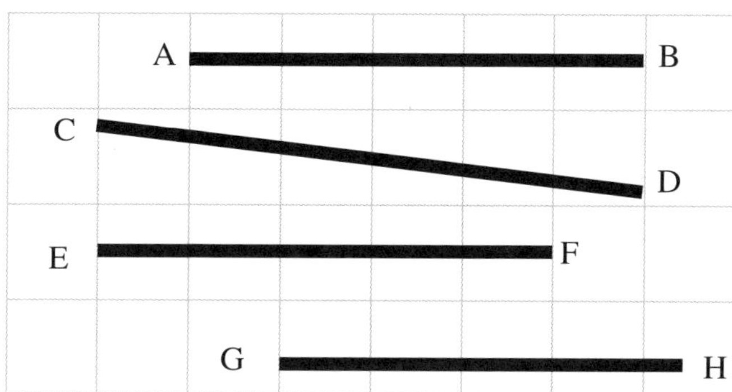

a) Line AB is longer than line ____________
b) Line AB is as long as line ____________
c) Line AB is shorter than line ____________

2 What is the length of each line?

a)

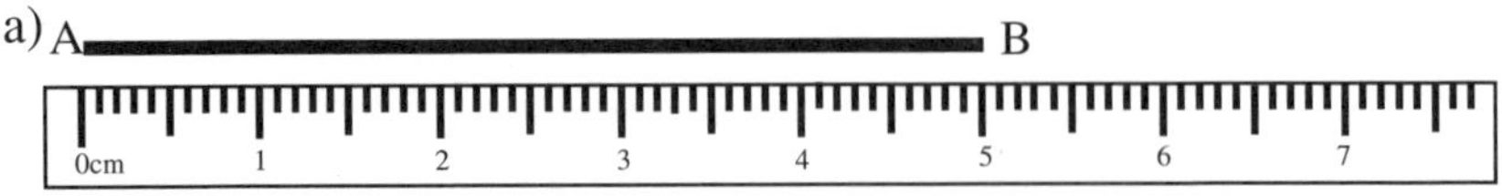

Length of line AB is ____________

b)

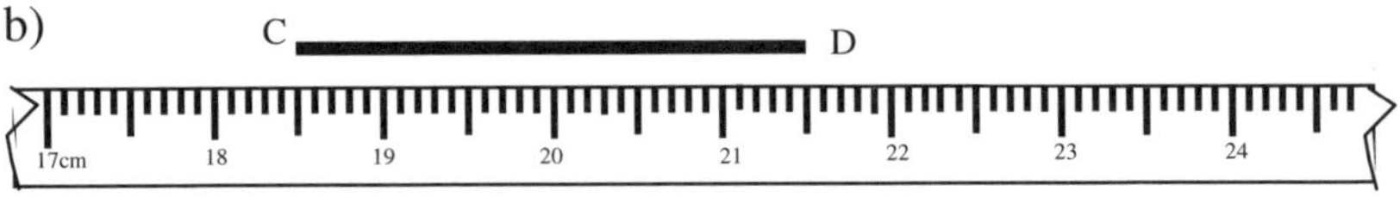

Length of line CD is ____________

3 Salma measures the width of the door using these different objects:

(i) paper clips (ii) matches (iii) cm cubes

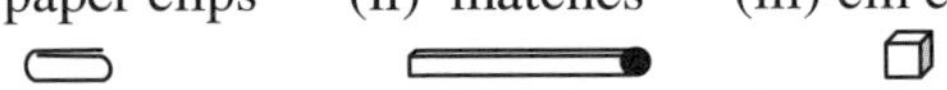

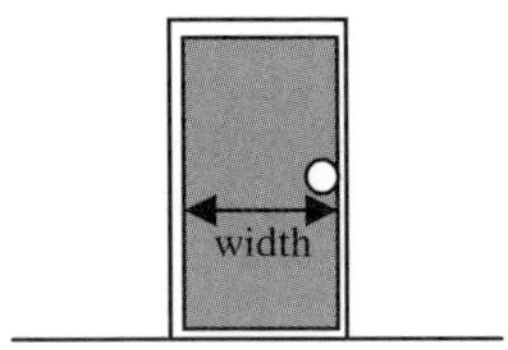

She gets the answers 95, 38 and 19.
Which number goes with each object?

a) Number of paper clips used: ____________

b) Number of matches used: ____________

c) Number of cm cubes used: ____________

4 Circle the shape which has the same <u>perimeter</u> as Shape A.

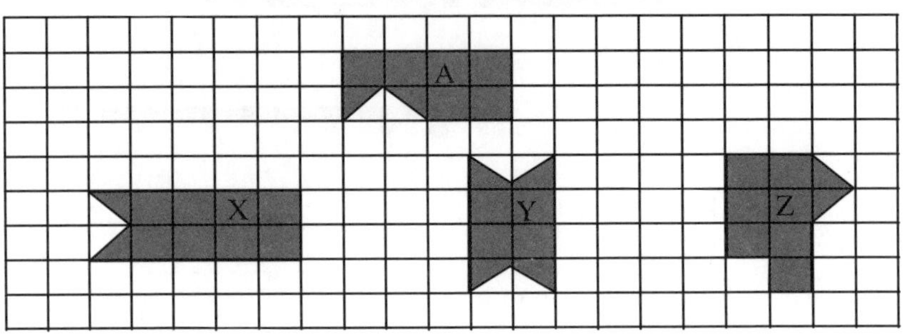

5 Find the <u>area</u> of each shape in square centimetres.

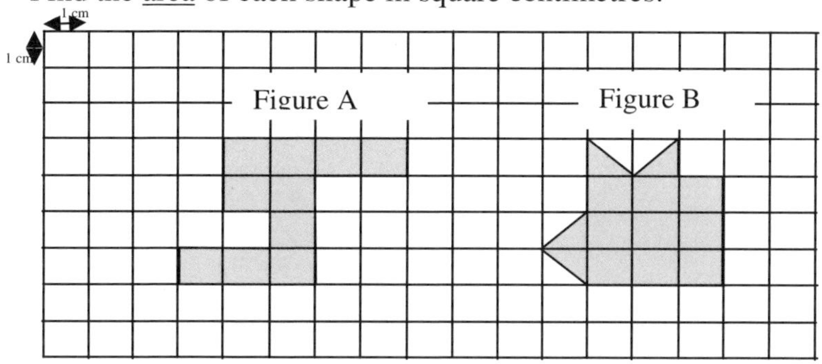

Area of Figure A = ____________ cm^2

Area of Figure B = ____________ cm^2

6 Fill in the blanks

a) 2650 g = ________ kg ________ g

b) 405 cm = ________ m ________ cm

c) 7400 *ml* = ________ *l* ________ *ml*

d) 3 m 20 cm = ________ cm

e) 6 *l* 507 *ml* = ________ *ml*

7 Shade the lightest mass.

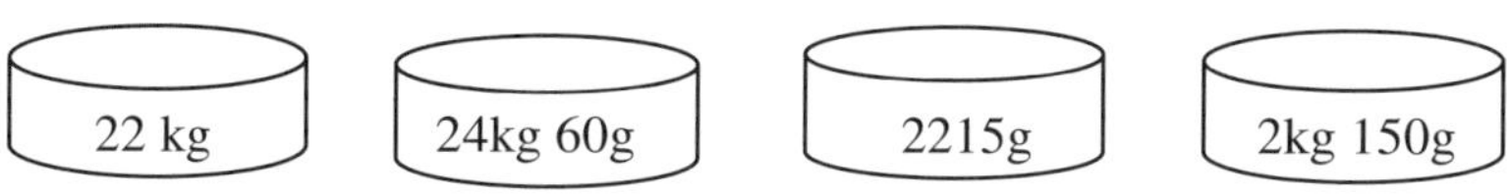

8 Fill in the blanks.

a) $37.20 = ____________ ¢

b) 3665¢ = $ ____________

Part II

9 The packets of rice have the same mass.
Find the mass of one packet of rice.

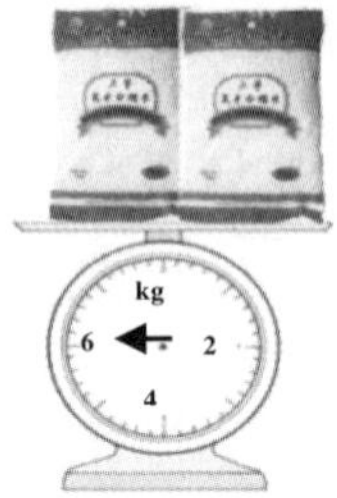

[kg]

---

10 A bamboo stick is 200 cm long.
A lamp pole is 3 m longer.
How long is the lamp pole?

[cm]

---

11 A pail holds 8 litres of water.
A tank holds 48 litres of water.
How many pails of water are needed to fill the tank?

[pails]

---

12 Lily has $57.80. Siti has $2 more than Lily.
How much does Siti have?

[$]

---

13 Ahmad has $88.65. Meng has $28.60.
a) Who has more money?
b) How much more?

[a)]

[b)$]

---

14 Joshua has $16.80. He has 60 cents less than Tim.
a) How much does Tim have?
b) How much money do they have altogether?

[a)$]

[b)$]

Appendix D4

Geometry

This test is based on the following content from the primary school mathematics syllabus (Ministry of Education, 2006) in Singapore.

Primary 1	Primary 2	Primary 3
▪ basic shapes (rectangle, square, circle, triangle) - identify, name, describe and classify shapes ▪ patterns - make/complete patterns with 2D cutouts - make / complete patterns with 3D models	▪ 2D & 3D figures (include semicircle and quarter circle) - identify, name, describe - identify the basic shapes that made up a given figure - from 2D figures with the basic shapes - form 3D solids - copy figures on grid paper ▪ patterns - make/complete patterns according to one/two of the attributes: shape, size, orientation &colour ▪ line, curve and surface - identify lines and curves - identify flat faces of a 3D object	▪ perpendicular and parallel lines - identify & name - draw on square grids ▪ angles - illustrate as amount of turning - identify angles in 2D and 3D objects - identify angles in 2D figures - identify right angles, angle greater / smaller than a right angles

1 Bala chose a shape that is made up of 3 straight lines and one curved line.

Figure 1 Figure 2 Figure 3 Figure 4

---

2 Circle the name of the shape that has four sides and four corners.

A triangle A circle A half circle A rectangle

---

3 Circle the letter(s) below that has a pair of parallel lines.

T K H X

---

4 Which lines in the figure below are perpendicular?
Draw along these perpendicular lines.

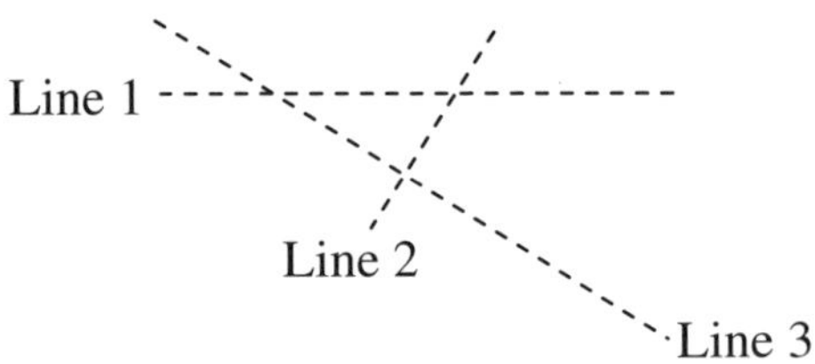

---

5 Circle the two equal angles.

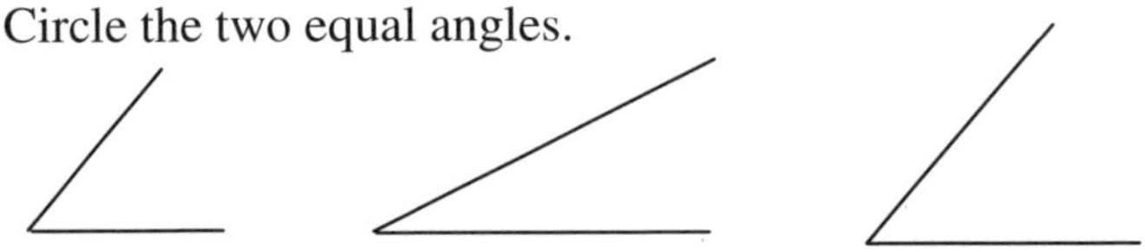

Figure 1 Figure 2 Figure 3

---

6 Write the letter A on the angle that is <u>larger</u> than a right angle.
Write the letter B on the angle that is <u>smaller</u> than a right angle.

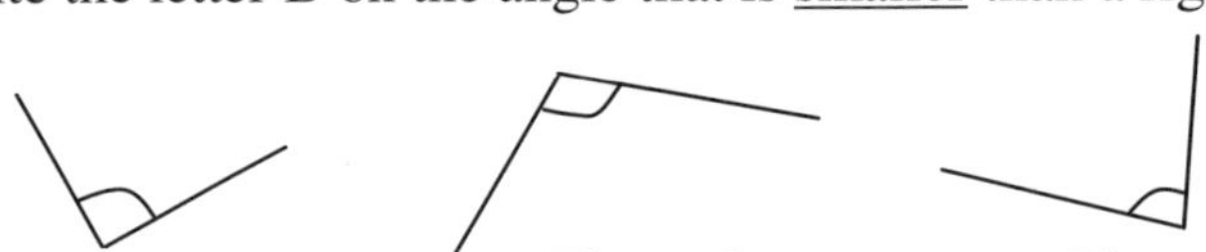

Figure 1 Figure 2 Figure 3

7 Look at the pattern

Circle the missing shape.

---

8 Look at the pattern

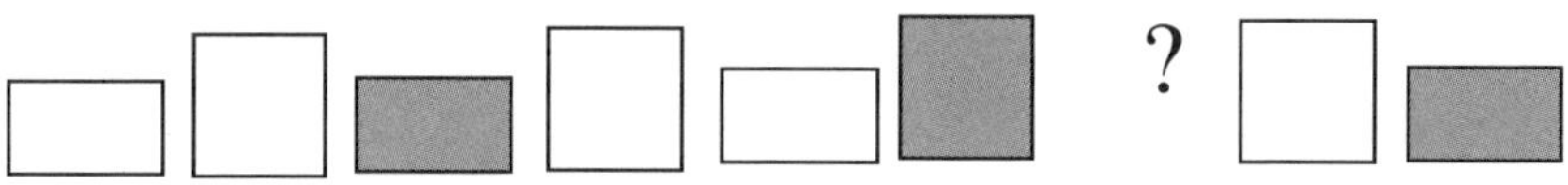

Circle the missing shape.

---

9 Peter built this structure. He used,

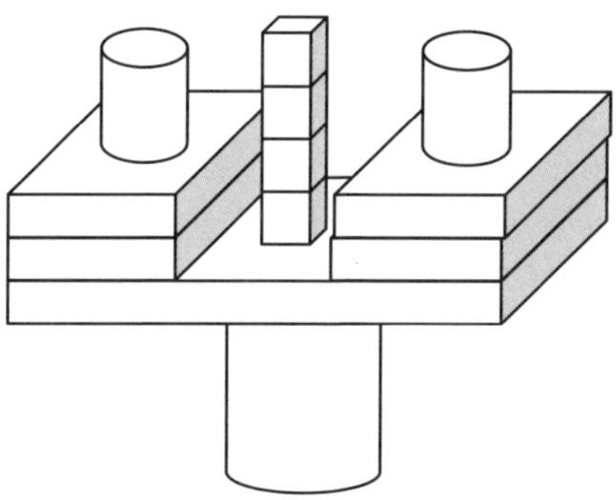

3	cylinders
	cubes
	cuboids

Appendix D5

Data analysis

This test is based on the following content from the primary school mathematics syllabus (Ministry of Education, 2006) in Singapore.

Primary 1	Primary 2	Primary 3
▪ picture graphs	▪ picture graphs with scales	▪ bar graphs

1 In P3A, every pupil has a pet.
Look at the graph below.

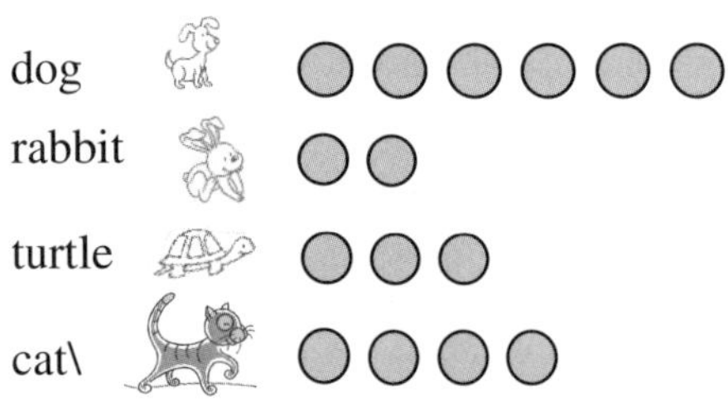

Each ○ stands for 2 P3A pupils

a) How many pupils have dogs?
b) How many more pupils own dogs than cats?
c) How many different pets do P3A pupils have?
d) How many pupils are there in P3A?

---

2 The graph shows the number of birthday cards Jane received from different countries. She received 6 cards from Australia.
Draw the missing bar in the graph to represent cards that she received from Australia.

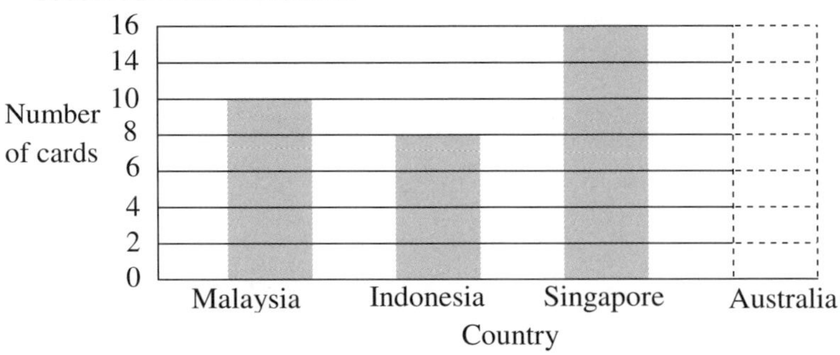

3 The graph shows the favourite M&M's colours of pupils from P4A

Favourite Colour of M&M's candies

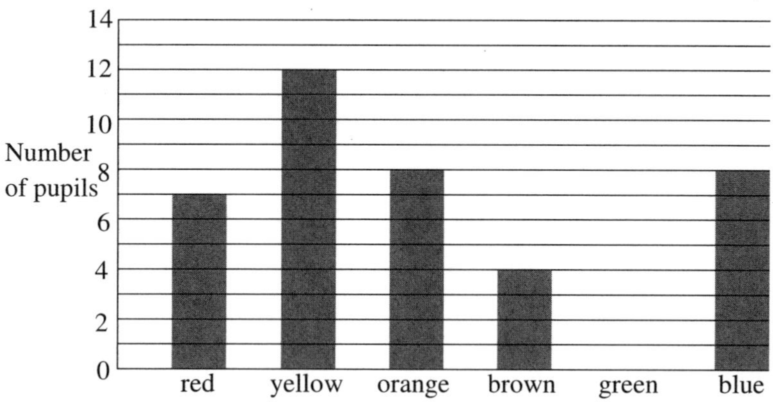

a) How many different colours of M&M's candies are there?
b) Which is the most popular colour?
c) How many pupils like green M&M's?
d) How many more pupils like orange M&M's than red M&M's?

---

4 This graph shows the scores obtained by 4 boys in a competition. Helmi is in the first place, Bala is in the third place. Draw a bar to represent the scores of Bala.

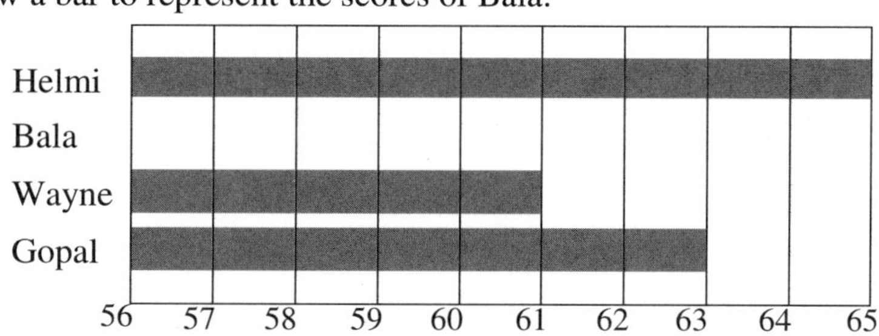

Appendix E

Pupil Profile Questionnaire

Name: ______________________ Class: __________

School: ______________________ Gender: __________

1. Race:
   Chinese Indian
   Malay Others: ______________(Please specify)

2. I live in a:
   3-room HDB flat Executive Maisonette/Apartment/ HUDC Flats
   4-room HDB flat Condominium/Private Apartment
   5-room HDB flat Landed Property: Terrace / Semi-detached / Bungalow
   Others: ________________________________(Please specify)

3a. Father's highest educational level:
   No formal schooling Vocational Institute /Institute of Education
   Primary Polytechnic
   Secondary University
   Junior College

3b. My father works as a ________________________________.

4a. Mother's highest educational level:
   No formal schooling Vocational Institute/Institute of Education
   Primary Polytechnic
   Secondary University
   Junior College

4b. My mother works as a ________________________________.

5. Most frequently spoken language at home is:

English	Tamil	
Mandarin	Dialects: ____________________	(Please specify)
Malay	Others: ____________________	(Please specify)

6. I have my own study table at home. Yes No

7. I have a computer at home. Yes No

8. Time spent on school mathematics homework per day:

None	About one hour
About two hours	More than two hours

9. Who helps you when you have problems with mathematics homework?

Father	No one
Mother	Tutor
Sibling	Others: ______________(Please specify)

10. Tuition in mathematics outside school

No tuition	Private one-to-one tuition
Private group tuition	Tuition class organised by community groups: SINDA/MENDAKI/CDAC.

Others : ____________________________________(Please specify)

Thank you for your participation

Appendix F

Pupil Behaviour Questionnaire

Dear P4 mathematics teacher,
Please help to fill out this questionnaire by circling the appropriate response for the named pupil in your class. Thank you.

Name of pupil: ______________________________

School: ______________________ Class: __________

1	Is the pupil attentive when you are teaching?	*Always*	*Sometimes*	*Seldom*	*Never*
2	Is the pupil on task during groupwork?	*Always*	*Sometimes*	*Seldom*	*Never*
3	Is the pupil on task during individual seatwork?	*Always*	*Sometimes*	*Seldom*	*Never*
4	Does the pupil seek help from the teacher when in doubt?	*Always*	*Sometimes*	*Seldom*	*Never*
5	Does the pupil complete his/her classwork on time?	*Always*	*Sometimes*	*Seldom*	*Never*
6	Does the pupil give his/her best when doing work in class?	*Always*	*Sometimes*	*Seldom*	*Never*
7	Is the pupil neat and organised in his/her work in mathematics?			*Yes*	*No*

8	Is the pupil a good listener?		*Yes*	*No*
9	Is the pupil easily distracted?		*Yes*	*No*
10	Is the pupil motivated to learn?		*Yes*	*No*
11	Is the pupil often away from school without valid reasons?		*Yes*	*No*
12	How is the pupil's proficiency in English Language?	*Good*	*Average*	*Poor*

13 Any other information you feel would help us to understand the pupil's mathematical attainment.

Thank you for your cooperation

Appendix G

Pupil Maths Learning Questionnaire

Name: ______________________________ Class: __________

School: ____

1. Do your parents check your maths work?

No
Yes, two or three times a week
Yes, everyday
Yes, only when they are free
Yes, once a week

2. Who help you at home when you have difficulties in your maths homework?

Parents
Brother and/or sister
Aunty / Uncle
Grandparents
Friends
Tutor

3. In addition to your homework, do you have extra work in maths at home?

No
Yes, work given by my tutor
Yes, work given by my parents
Yes, assessment books that I bought

4. Put a tick in the boxes to show what you think.

Pedagogy		Yes	No
a)	My maths teacher always teaches maths in the same way. She stands in front and teaches.		
b)	We do a lot of group activities in our maths lessons.		
c)	I have maths homework every day.		
d)	Some maths homework is fun to do.		

5. Shade a face to show what you think.

Perception			
a) Maths is easy.			
b) Maths is important.			
c) Maths is fun.			
d) Maths is useful in daily life.			

6. Put a tick in the boxes to show what you think.

Self-Confidence		Always	Sometimes	Never
a)	I am good at maths.			
b)	I give up when a problem looks difficult.			
c)	I make careless mistakes.			
d)	I get stuck when solving maths problems.			

7. Put a tick in the boxes to show what you think.

Learning Preference		Yes	No
a)	I like to solve simple maths problems more than the difficult ones.		
b)	I like to follow rules/steps when I do maths.		
c)	I like to learn maths by doing activities.		
d)	I like to know how maths is used in my daily life.		
e)	I like to work in groups during maths lessons.		
f)	I like to see maths charts and posters in my classroom.		
g)	Stories help me learn maths better.		
h)	Diagrams and figures help me understand a problem or idea better.		
i)	I enjoy working on maths worksheets.		

Thank you for your participation

Appendix H

Pupil Interview Prompts

Pupils will be interviewed in groups of 4-5.
Estimated time of the interview session: approximately 45-60 minutes

1. Pedagogy

Picture A  Teacher-led whole class instruction	Picture B 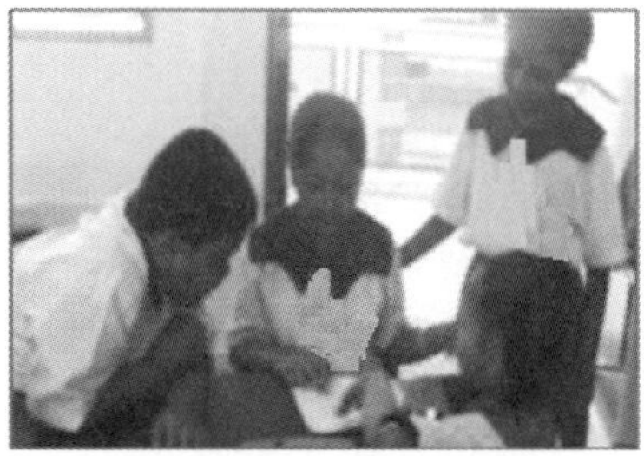 Group work (pupils working on tasks without manipulatives)
Picture C  Individual working on task with manipulatives	Picture D  Group work (pupils working on tasks with manipulatives)

- Show the four pictures and explain to the pupils that:
  i) In Picture A, a teacher stands in front of the board and teaches mathematics;
  ii) In Picture B, the pupils talk about and do mathematics in a group in their mathematics lesson;
  iii) In Picture C, a pupil carries out a mathematical activity alone in her mathematics lesson; and
  iv) In Picture D, pupils work in a group and carry out a mathematical activity in their mathematics lesson.

- Ask (Try to get each child to contribute. Whenever possible, mention the child's name)
  1. Which picture shows the way your maths teacher usually teaches you in class?
  2. Which class do you want to be in? Why?
  3. Which class don't you want to be in? Why not?

2. Homework
   - Show the three types of homework on money; Homework A, Homework B and Homework C. *[Pupils have already done the topic money at school prior to the interview.]*

Homework A

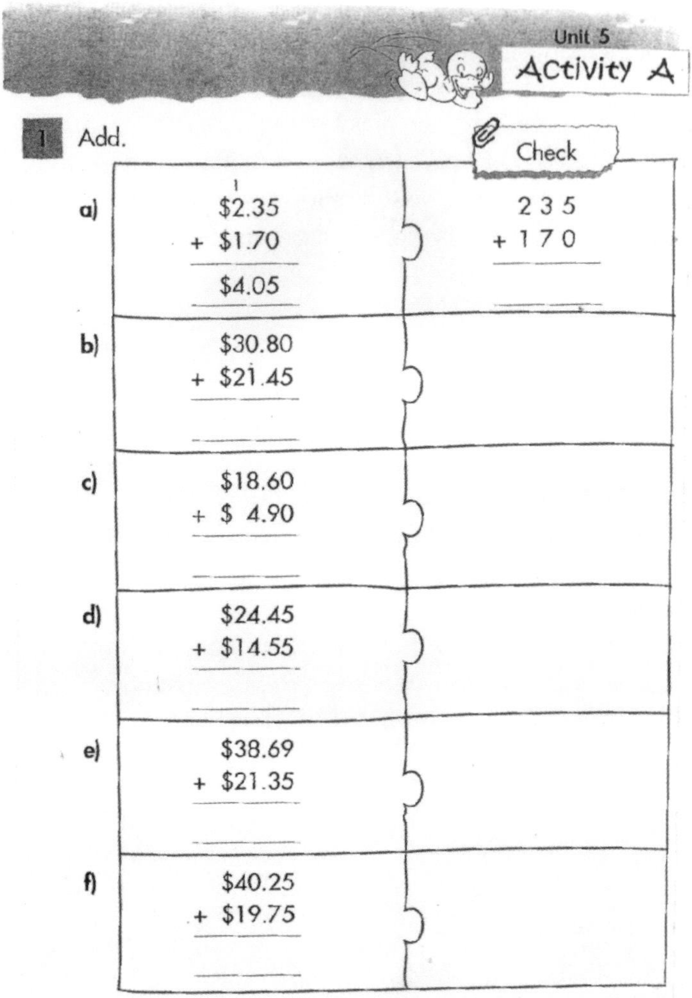

Unit 5

Activity A

1 Add.

		Check
a)	1 $2.35 + $1.70 $4.05	2 3 5 + 1 7 0
b)	$30.80 + $21.45	
c)	$18.60 + $ 4.90	
d)	$24.45 + $14.55	
e)	$38.69 + $21.35	
f)	$40.25 + $19.75	

A typical piece of exercise reinforcing algorithms taught in class.

Homework B

1 Mrs Lim bought some groceries from the supermarket.
The groceries cost $87.35.
If Mrs Lim gave the cashier $90, how much change would she get?

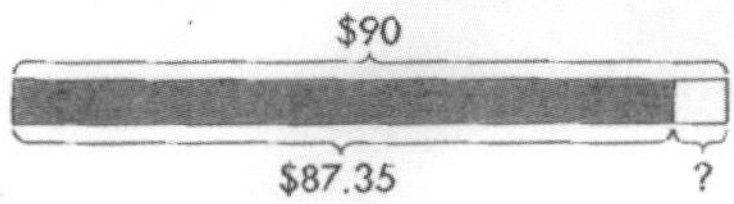

2 Susan saved $8.50 this week.
She saved $4.85 more last week than this week.
How much did she save last week?

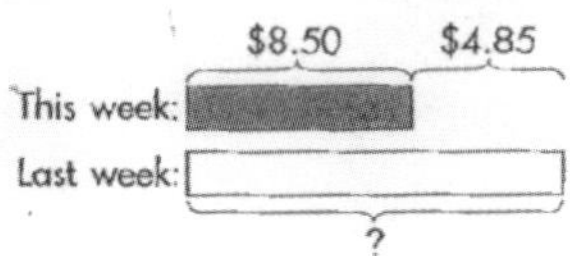

Multi-step word problems that may require the drawing of models and writing of statements to arrive at a solution

Homework C

Unit 5

Activity C

1 **a)** Circle any two types of food which you like to eat.

b) Find their total cost.

In the box, paste the notes and coins that make this amount.

Non-routine activity involving manipulatives to represent mathematical concepts.

Manipulatives for Homework C

- Ask (Try to get each child to contribute. Whenever possible, mention the child's name)
  1. Which type of homework do you usually get in the maths classrooms?

     Do you sometimes also get the other types of homework?
  2. Which type of homework do you like best? Why do you like this type of homework?
  3. Which type(s) of homework don't you like? Why not?

3. Self

We have talked about what happens in the mathematics classroom and your homework, let's talk about you.

1. Which part(s) of school maths don't you like most? Why is it so?
2. Which part(s) of school maths do you like most? Why is it so?
3. Do you think you are good or not good in maths? How can you tell?
4. Do you think you can do better in maths? Why or why not?
5. How can you be better in maths? What/who can help you to do better in maths? How do they help you? Do they use things to help you?

Appendix I

Pupil Journal

Appendix J

Lesson Observation Analytical Questions

Our lesson observations were guided by the following main analytical questions that resulted from our theoretical framework. The questions are:

Item	Analytical Question
1a	What was the focus of the lesson - creation or dissemination of knowledge?
1b	What was the instructional sequence of the lesson like?
2	Did the teacher tailor instruction to meet the needs of different learners?
3	What were the characteristics of mathematical tasks used in the lesson?
4a	Did the teacher ask pupils questions (both open and closed)?
4b	Did the teacher praise pupils for their responses?
4c	Did the teacher build on the pupils' responses (both correct and incorrect) to explain mathematical concepts?
4d	Did the teacher encourage pupils to ask questions?
4e	Did the pupils ask the teacher any questions to clarify their doubts?
4f	Did the pupils ask their peers any questions to clarify their ideas?

Contributing Authors

CHANG Suo Hui is a teaching fellow with the Mathematics Education Academic Group at the National Institute of Education, Singapore. Her teaching duties include pre-service modules of Teaching of Primary Mathematics I, II and III with the Degree, Diploma and Postgraduate student teachers. Prior to her stint with NIE, she taught in a Primary School for ten years and had served as the Head of Department for Physical Education. Her research interest includes examining the performance of low attainers in different mathematical concepts and skills, in particular, their learning difficulties, as well as various teaching instructions that aid learning for this group of pupils. In addition, she is currently exploring the possibility of harnessing computer games in the incidental learning of primary mathematics.

FOONG Pui Yee is an Associate Professor of Mathematics Education at the National Institute of Education, Singapore. She obtained her PhD from Monash University, Australia, with a study on a metacognitive-heuristic approach to mathematical problem solving. Her teaching and research are focused on the area of primary school mathematics. She taught courses in the pre-service, in-service and graduate programmes at the Institute. She also supervised Master and PhD candidates in the areas of mathematical problem solving, attitudes and learning difficulties in primary mathematics.

Masura GHANI is a research associate at the Centre of Research and Pedagogy Practices in the National Institute of Education, Singapore. She has 12 years of teaching experience in Singapore and was a subject head of mathematics prior to her stint with NIE. She has been credited with Outstanding Contribution Awards as well as the Caring Teacher Award during her teaching service in Singapore. She has also taught in two schools in Brisbane, Australia, while pursuing her master degree in 2007.

Berinderjeet KAUR, is a Professor of Mathematics Education and Head of the Centre for International Comparative Studies (CICS) at the National Institute of Education in Singapore. She obtained her PhD from Monash University, Australia, with a study of children's knowledge and strategies in mathematical problem solving. Her primary research interests are in the area of classroom pedagogy of mathematics teachers and comparative studies in mathematics education. She has been involved in numerous international studies of Mathematics Education and is the Mathematics Consultant to TIMSS 2011. She is the principal investigator (Singapore) of the Learner's Perspective Study (LPS) helmed by Professor David Clarke of the University of Melbourne. As the President of the Association of Mathematics Educators (AME) from 2004-2010, she has also been actively involved in the professional development of mathematics teachers in Singapore and is the founding chairperson of the Mathematics Teachers Conferences that started in 2005. She is also the founding editor of the AME Yearbook series that started in 2009. On Singapore's 41st National Day in 2006, she was awarded the Public Administration Medal by the President of Singapore.

KOAY Phong Lee is a Senior Lecturer of Mathematics Education at the National Institute of Education in Singapore. She obtained her PhD at University of Queensland with a study on how primary school children learn the algorithms from the information processing perspectives. Her research interests continue to be in the area of primary school mathematics. She is also involved in the writing of primary mathematics textbooks used in Singapore.